U0186230

食物小传

威士忌

Whiskey

A Global History

〔美国〕凯文·R. 科萨 著

吴德煊 译

北京联合出版公司
Beijing United Publishing Co.,Ltd.

目录

　　威士忌是迷人的。世界上的威士忌品牌有数百个，但是几乎没有哪两款威士忌尝起来是一样的。雅柏和拉弗格都是单一麦芽苏格兰威士忌，生产它们的两家酿酒厂都位于苏格兰西海岸的艾雷岛上，相隔仅一箭之遥，但两家所产威士忌的风味却大相径庭。

　　每一款威士忌都可以反映出其酿酒原料的质量、酿酒设备的性能和酿酒师的技艺。时间的流逝会对所有这些酿酒因素产生或大或小的影响：酿酒方法会不断调整，水、土壤和谷物会发生变化，蒸馏厂会变得陈旧，而酿酒师也会老去，直至被取而代之。新威士忌品牌诞生，老威士忌品牌倒下，还有一些品牌涅槃重生。因此，2010 年生产的布鲁克莱迪苏格兰威士忌的味道和 1950 年产的并不会完全一样。对于具有好奇心的人来说，不断演变的威士忌也为他们提供了无尽的探索机会。

　　我当时也醉心于写作此书，因为威士忌并不单单是一种饮品，它还是一种政治、经济和文化现象。威士忌的起源可以追溯到 15 世纪之前，它最初只是一种默默无闻、经常给人带来喉咙灼烧感的饮品，酿造方式粗糙，只有不列颠群岛上贫困村落的住户和农民才喝。如今，威士忌的种类和品牌多到令人惊叹，其消费者中有穷人也有富人，有乡村居民也有城市居民。威士忌的生产也已远渡重洋，遍及德国、日本、巴基斯坦和新西兰等地。现在威士忌酿造既是一门技术，又是一门艺术。

这个世界真的需要多一本关于威士忌的书吗？答案是肯定的。

大多数关于威士忌的书籍要么是品鉴指南，要么只介绍单个产酒国的威士忌种类（如爱尔兰威士忌、苏格兰威士忌、美国波本威士忌等），而后者常常侧重于讲述独立蒸馏师和品牌的故事。

虽然这种写作思路兼具启发性和趣味性，但是它错过了更大的创作图景，也忽视了各国威士忌发展历程中惊人的相似之处：政府为对威士忌进行合理监管和征税制定政策所经历的挣扎，时而被这些政策激起的暴力回应，威士忌生产的繁荣、萧条和工业化之路，威士忌的政治化，威士忌对民族文化的渗透，还有由其引发的道德上的强烈抵制。此外，由于如今所有威士忌都是威士忌世界版图的一部分，并且都在为争夺消费者而竞争，所以将全球各地的威士忌放在一起研究是有意义的。

我还必须补充一点，太多威士忌书的作者都对他们的题材爱到无可救药了。他们不加批判地重复威士忌生产商的高谈阔论和公关公司编造的无稽之谈，并且还描绘了一幅其乐融融的景象：精明的老前辈与有着强大后盾的年轻人都在用"老法子"酿造威士忌。我曾经参观过一些蒸馏厂和威士忌小镇，很多地方都风景秀丽，那些地方的人也令人难忘。威士忌一直在诗篇和歌曲中被广为传颂，并不是没有原因的。

然而，威士忌世界同样存在着世俗和丑陋的一面。首先，威士忌是一门生意。一些业余爱好者可能会制作极小批量的威士忌供自己饮用，但大多数威士忌都产于计算机化的工厂，那里有专业人士密切关注着收益情况。此外，威士忌是一种烈性饮品，很多人都无法驾驭。威士忌的误用和滥用引发斗殴、毁掉家庭，甚至让许多人付出了生命的代价。

这是个疯狂的故事，要从黑暗时代的不列颠群岛一直讲到 21 世纪的新世界，情节涵盖了政治动荡、技术革命、犯罪集团、道德抵制和全球化；故事角色阵容丰富，炼金术士、骗子、怪人、诗人、政治家、传教士、科学家和无数只想享用一杯烈酒的普通人都在其中。

[第一章]

起源：从种子到烈酒

whisky 与 whiskey

关于威士忌有很多毫无根据的传言，比如威士忌的英文 whisky 和 whiskey 的"正确"用法。我听到过有人坚持认为 whiskey 这个词特指美国生产的威士忌，但这种说法是错误的，只需要看一看美国生产的老林头肯塔基纯波本威士忌酒瓶就知道了——它的酒标用的就是 whisky。

一般来说，在加拿大、英格兰和苏格兰，人们更倾向于使用 whisky 这个拼写，而在爱尔兰和美国则是 whiskey。但是也有许多不符合这些习惯的例外情况存在。

本书中出现的 whiskey 是个笼统的泛称，whisky 和 whiskey 都被涵盖在内；不带字母 e 的 whisky 只用于提及苏格兰和加拿大威士忌，或用于特定的品牌，比如乔治迪科尔 12 号田纳西威士忌。

威士忌的定义

简单来说，威士忌是一种由发酵谷物经过蒸馏和橡木桶陈酿而成的酒精饮料。"谷物"一词指的是人类耕种用来食用的会结籽的禾本科植物，例如大麦、玉米、燕麦、黑麦、小麦等等。橡木桶陈酿是威士忌定义中一个必不可少的部分，因为正是这个过程赋予了威士忌颜色（从稻草黄到深棕色）和风味（如香草风味）。综合来看，这两个特点将威士忌和其他蒸馏烈酒区别开来，比如干邑（由发酵葡

19世纪70年代的一则老乌鸦"威士忌"广告

萄汁蒸馏而成）和伏特加（可以由谷物、甜菜等几乎所有能发酵的原料蒸馏而成，但不经过橡木桶陈酿）。

和其他烈酒一样，现在大多数威士忌装瓶时酒精纯度是80至86 proof[1]，即酒精度40%至43%。但并不是所有威士忌的酒精含量都在这个范围，比如乔治·T.斯塔格肯塔基纯波本威士忌，它在装瓶时的酒精纯度就是让人头晕眼花的142.7 proof（酒精度71.35%）。

威士忌的酿造

酿造威士忌必须经过两个重要步骤：发酵和蒸馏。

简单来说，发酵就是用糖、酵母和水产生酒精。啤酒和葡萄酒就是分别由谷物和葡萄发酵而来的。就葡萄酒来

---

[1] 美制酒度用酒精纯度 (proof) 表示，1 个酒精纯度等于 0.5% 的酒精含量。多见于美国酒标。——本书注释，除特别说明，均为译注

说，将葡萄压榨出甜甜的葡萄汁，葡萄自身带有的酵母或酿酒师添加的酵母在糖类物质上饱餐一顿之后便会产生二氧化碳气体和酒精。至于啤酒，还需要多一个步骤：在酵母发挥作用之前，谷物必须经过热水煮制，以释放糖分。

蒸馏，就是将液体加热至其沸点，收集蒸汽，经过冷凝将蒸汽重新转化为液体的过程。所以，我们可以把酿造威士忌理解为将谷物和水发酵成啤酒，再将啤酒蒸馏成烈酒。

这么看来，制作威士忌很简单，不是吗？先做出啤酒，再将啤酒煮沸，把醉人的蒸汽收集起来，最后冷凝成烈酒——瞧，这就成了！就算再蠢的人也能照做！

但是呢，答案是也不是。做出难喝、烧喉的劣质威士忌确实挺简单，但是要想酿出美味、令人感官愉悦的优质威士忌，却是件极具挑战性的事。很多人尝试去做了，然

大麦——威士忌的核心原料

而成功的人却寥寥无几。

　　这是因为酿酒过程中的许多因素对最终成品的味道都有着巨大的影响，影响可能是好，也可能是坏。这些因素包括酿酒用的水的特性，酵母和谷物的质量、数量和种类，蒸馏器的形状、结构和工作原理，以及威士忌的蒸馏次数。蒸馏之后，熟成过程也会带来无数的变化因素，比如用多少时间来陈酿威士忌，陈酿用的橡木桶有什么特征。橡木桶特征具体包括橡木桶的大小，木材类型和内部状况（如新旧程度，烘烤程度，是否曾用于存放雪莉酒、波本威士忌或其他酒精饮料），以及仓库中橡木桶存放区域的微气候。此外，威士忌的酿造过程本身就具有挑战性。它是化学与手工艺的天作之合，包含了四个阶段：谷物制备、发酵、蒸馏、熟成和装瓶。整个过程都需要格外用心对待，才能酿造出口感极佳的威士忌。

　　每家蒸馏厂都有各自的酿酒方式和方法。以下是威士忌生产过程的精简概述。

谷物制备

　　为了从种子中获得烈酒，威士忌酿酒师必须挑选富含淀粉的硬粒种子，再将其转化为适合发酵和蒸馏的含糖物质。

　　要了解其中原理，我们需要先解决一个基本的问题：什么是种子？简单地说，种子就是休眠的幼苗。其内部是渴望发芽的胚，外部则是种皮，即一层对胚起保护作用并为其提供营养的淀粉层。有了适宜的环境（光照和温度）和水分，种子就会苏醒并开始生长，其产生的酶（细胞解糖酶和淀粉糖化酶）有助于将种皮转化为碳水化合物（糊精）和糖（麦芽糖），为种子的生长提供能量。然而，威士忌酿酒师却对这糖另有打算。于是，他会促使种子生产出糖分，但之后却不让糖分为种子所用。

为了达到这样的效果，谷物制备大致分为三个阶段：麦芽制造、研磨、糖化。

麦芽制造

麦芽制造阶段是为了得到"麦芽"。谷物种子在被研磨和糖化之后，可以进而被发酵和蒸馏成威士忌。麦芽制造包含三个步骤——浸麦、发芽和烘干。

威士忌酿酒师会将谷物种子放入装满水的浸麦槽中连续浸泡几天。浸泡的时间长短至关重要：如果种子吸收的水分不足，种子便无法完全生长；如果吸收水分过多，则会变成糊状物。浸泡完成后，酿酒师会将谷粒转移到大号容器中，种子便开始发芽（生长）。此时它们的胚会苏醒，其新陈代谢的过程也就开始了。

一旦种子生长到特定的阶段，威士忌酿酒师就会通过烘干（又称烘烤）来打断发芽的进程。热气在种子之间穿梭，这个过程持续一到两天。看到过或参观过威士忌蒸馏厂的人可能会对宝塔式的小房顶印象深刻，而这种造型的

苏格兰艾雷岛上拉弗格蒸馏厂的发芽制造间

屋顶可以将60摄氏度（140华氏度）的热气排出蒸馏厂。怎样加热烘干气体本不是什么特别有趣的话题，但是涉及苏格兰威士忌就不一样了。制作苏格兰威士忌时，加热空气用的炉子里面烧的是砖块状的泥煤和腐烂的植物。泥煤燃烧的时候会散发出烟味、海草味和其他强烈的气味，这些气味会附着在被烘干的大麦种子上，这就是苏格兰威士忌闻起来和尝起来常常带有烟熏味的原因。

研　磨

经过制麦的谷粒，也就是麦芽，现在已经准备好被研磨了。为了理解研磨的重要性，我们有必要回顾下面包是怎么制作的。如果你直接将水和酵母加到一碗黑麦或小麦种子里，碗里不会起什么反应。但是，如果你先把碗里的黑麦或小麦种子磨成粉末，然后再加入水和酵母，那么你就能得到一个面团，发酵膨胀后可烘烤成面包。同样的原理大致上也适用于威士忌。威士忌酿酒师会将发芽的种子送进一个修整机，把须根从种子上去掉，再将种子倒进研磨机磨碎成粉。

糖 化

到这个阶段，威士忌酿酒师就可以提取出种子里的糖分，随后将其转化为酒精了。这个过程是通过糖化来实现的——将碾碎的谷物倒进糖化锅（一种装有热水的大水箱，里面耙子状的机械臂会对稀粥状混合物进行搅拌），经过至少30分钟的搅拌以后，将水从糖化锅底部的开口排出，然后再倾倒入更多（且可能更热）的水。这一过程可能会重复一至三次。糖化过程会激活淀粉糖化酶，将麦芽中的淀粉和糊精转化为糖。糖化的效果相当显著，原本富含淀粉、难以下咽的糊状物变成了带有甜味的可饮用液体。

麦芽汁是麦芽浆中最有价值的部分，它从糖化锅中排出之后会流经热交换器。热气腾腾的甜汤在热交换器里冷却下来，再流入发酵槽（一个用来进行发酵步骤的大池子）。

苏格兰艾雷岛的布鲁克莱迪蒸馏厂中,热水正被倒入糖化锅。

发 酵

这场发酵表演的主角便是酵母,一种来自真菌王国的单细胞微生物。酵母的一生简简单单:无意识地进食,繁殖,然后死亡。到目前为止,科学家们已经鉴别出了 1500 多种不同的酵母。

有一种酵母特别受制酒者青睐,那就是酿酒酵母,拉丁学名为 *Saccharomyces cerevisiae*,翻译过来意为"啤酒的糖霉菌"。这种酵母依靠糖分生长,因为糖分为它提供了无性繁殖所需的动力——新酵母直接从成熟酵母身上出芽,这有点像从宙斯头颅中一跃而出的雅典娜①。在酿酒酵母的作用下，糖类物质便会转化为酒精、二氧化碳气体和同系物；最后这个术语是一个概括性的词，指的是一众酸类和脂类物质。同系物中混合了多种成分，有些能够为威士忌提供理想的风味，有些则会起负面作用。

和其他生物一样，酵母只能在一定的温度范围内生长，这个范围大概是 10 至 37.8 摄氏度（约 50 至 100 华氏度）。过低的温度会使酵母失去活力，而过高的温度对酵母来说是一种折磨。直接将酿酒酵母加入到滚烫的麦芽汁中，无异于将其立即处死，徒留这一大罐谷物和水，什么变化都不会发生。（这就是麦芽汁需要经过热交换器进行冷却的原因。）

所有的酵母都有不同的作用，所产生的同系物也不一样。因此，和啤酒或葡萄酒酿酒师一样，蒸馏师对选用的酿酒酵母株系是极为挑剔的。许多威士忌酿酒师会培养自家的酵母菌落，并且还会将多余的酵母菌落保存在远离蒸馏厂的地方，以防万一。每次发酵都使用完全相同的酵母株系，这是保证威士忌成品尝起来符合酿酒师预期、满足消费者期待的必要步骤。

---

① 希腊神话中，墨提斯是宙斯的第一任妻子。据传言，她的儿子将会推翻宙斯的统治，为了杜绝后患，宙斯便将怀孕的她吞进了肚里。但由于墨提斯是不死之神，胎儿也在宙斯肚中正常发育。最后，一个女孩从他的头颅中蹦出，而她就是雅典娜。

肯塔基州克莱蒙特区的金宾蒸馏厂中，麦芽汁正在发酵槽里进行发酵。

富含糖分的麦芽汁是酵母的天堂，在那里，酵母尽情饱餐，生生不息。发酵阶段这就开始了。发酵槽里的麦芽汁会不断冒泡泡和起泡沫，持续至少两天时间，直到麦芽汁的酒精含量达到 5% 至 10%。然后这场派对就到了散场的时候：酵母停止活动，进入假死状态。这场狂欢活动的成果就是得到了低酒精度的酒，也叫酒醪或蒸馏师的啤酒。

蒸　馏

尽管酒醪里含有酒精，但它仍然是有机物。几小时内，空气中的微生物就会侵入酒醪，引起腐烂。为了避免腐烂发生，威士忌酿酒师会尽快将酒醪转移到蒸馏器中进行蒸馏。有些酿酒师会将整桶酒醪倒进蒸馏器，而有些酿酒师只选取含水量最高的那部分，舍弃掉由酵母和谷物粉末组成的黏状物。这都取决于威士忌酿酒师的口味偏好——任何进入蒸馏器的物质都会影响蒸馏出来的酒液的风味。

虽然每个蒸馏器的形状和大小各有不同，但粗略来说，蒸馏器有两种类型：壶式蒸馏器和柱式蒸馏器（又称为连续式蒸馏器或科菲蒸馏器）。壶式蒸馏器看起来有点儿像一个巨大的铜葫芦，底部是球状，向上收窄成一个大幅度弯曲的天鹅颈。制作威士忌需要用到两到三个壶式蒸馏器。柱式蒸馏器高度至少有 30 英尺（9 米），由两个或两个以上的机械柱组成。壶式蒸馏器已存在了近千年，而柱式蒸馏器直到 19 世纪才出现。

每次蒸馏的时候，威士忌酿酒师会用蒸馏器加热之前倒入的酒醪。温度必须恰到好处——要高到足以将酒醪中的酒精蒸发，但又不能高到让酒醪中的水也蒸发。（威士忌厂商生产的是威士忌，不是蒸馏水。）酿酒师还需要注意控制火候，以免将酒醪烧焦而给蒸馏酒带来难闻的焦味儿。高温的酒精蒸汽不断上升，直到撞上"冷凝器"——这种冷却铜管（用于壶式蒸馏器）或铜板（用于柱式蒸馏器）可以将蒸汽冷凝为清澈的蒸馏烈酒。铜在蒸馏过程中起着关键作用。在对酒醪进行加热的过程中，一些无益的化合物，比如含硫化合物，会和铜发生反应形成油性混合物，有时也被叫作"脏东西"。这些混合物会被留在铜制蒸馏器里，所以不会影响到冷凝中的烈酒。

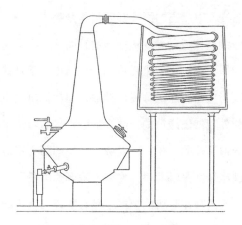

单个壶式蒸馏器示意图

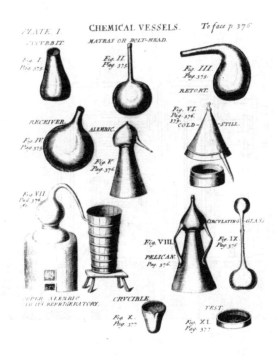

赫尔曼·布尔哈夫于 1727 年绘制的蒸馏器图解

蒸馏过程中产出的酒液并非都是好喝的。威士忌酿酒师必须仔细监控流动着的蒸馏烈酒，而这一步是通过观察从冷凝器里流出的酒液来完成的。仪器（通常是计算机化的）可以为威士忌酿酒师提供读数，但他们自己也会对酒的颜色和外观做出谨慎的判断。最先和最后蒸馏出的酒液都含有味道不佳的物质，所以威士忌酿酒师会将馏出液的酒头和酒尾送回蒸馏器里，将令人垂涎的中段酒导入酒精收集器。酿酒师在时机的把握上稍有不慎，就可能对威士忌产生负面影响，而这种影响要多年后才会被察觉。

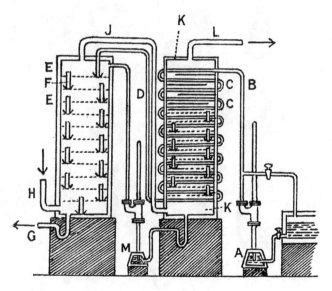

连续式科菲蒸馏器图解

熟成和装瓶

酒精收集器里的威士忌如水般清澈，且相当强劲——酒精纯度达140至160 proof（酒精度70%至80%）。通常情况下，威士忌酿酒师在将威士忌装入橡木桶之前，会通过加水来降低酒精度。不过，有些蒸馏厂也会出售"原桶强度"威士忌，意思是他们装瓶前未曾加水稀释酒精度。这样一来，我们才可以在市面上找到120 proof（酒精度60%）的康尼马拉原桶强度爱尔兰威士忌和其他原桶强度威士忌。

威士忌酿酒师和营销人员会经常大谈橡木桶陈酿的重要性，以及他们所用橡木桶的特质和存储环境。事实确如他们所说。威士忌的颜色和大约一半的风味都是由橡木桶陈酿这个步骤赋予的。威士忌一经装瓶，即使再发生变化，也是微乎其微了。

橡木桶有一系列不同的型号。大多数美国威士忌是在53加仑容量的橡木桶里陈酿的。苏格兰威士忌则会被装在50加仑的橡木桶、66加仑的猪头桶和132加仑的巴特桶里。桶的型号很重要。橡木桶越小，威士忌从橡木桶获取颜色和风味的速度就越快。通过一个简单的数学计算就能得知，在小橡木桶里，木材表面与威士忌酒液的接触比例要比在大橡木桶内的更高。例如，猪头桶里平均每升酒液接触到的木材表面面积，比巴特桶内的要多25%。

木材种类和桶陈过程会对威士忌产生极大的影响。有些威士忌会存储于用欧洲橡木制成的桶里，不过大多数威

克里尼利基蒸馏厂的壶式蒸馏器。该厂位于苏格兰萨瑟兰郡布罗拉村。

一位工作中的制桶工人

士忌是在美国橡木桶里进行陈酿的。威士忌酿酒师对生长于寒冷气候的橡木制成的桶情有独钟：它们生长得比较缓慢，所以木材的密度更高，而木材密度越高，做出来的橡木桶就越耐用，威士忌在桶中蒸发的量就越少。威士忌的蒸发可不是件小事，仅仅四年的时间就可能让一个橡木桶里的威士忌损失多达12加仑。如果乘以成百上千个橡木桶，你会发现一大笔钱就这样凭空蒸发。酿酒师将这种情况称为天使的份额。

　　不同类型的威士忌会使用不同条件的橡木桶来陈酿。美国政府的法规规定，波本威士忌必须经由未曾使用过的白橡木桶进行陈酿，这种桶的内部是被明火烘烤而炭化过的。爱尔兰和苏格兰的威士忌酒厂则经常选择曾陈酿过波本威士忌、马德拉酒或雪莉酒的橡木桶。

威士忌在橡木桶中陈酿时间的长短，取决于威士忌生产商的判断、经济效益和法律要求。威士忌生产商都想让自家酿的威士忌风味达到最佳。五年陈的威士忌通常会比两年陈的口感更好些。然而，所有产出的威士忌都面临着一个问题，就是陈酿带来的效果会越来越弱，甚至出现负面效果。因此，一款二十年陈的威士忌尝起来可能真的不如只将它陈酿十年的味道来得好。另外，出于财务层面的考虑，威士忌生产商也必须将酒推向市场出售。高年份威士忌卖出的价格确实经常比低年份的高，但是酒厂当前的经营成本也必须由厂商当即支付。各国通过制定法规条例，对特定类型的威士忌的最低陈酿时间做出了规定。美国规定波本威士忌的陈酿时间不少于两年；加拿大、爱尔兰和苏格兰则要求他们的威士忌陈酿时间不少于三年。

　　陈酿完成后，威士忌会被人们从橡木桶中取出，直接装入瓶中。而混合纯麦威士忌和调和威士忌却是此规则的例外。在这两种情况下，桶里的威士忌会被倒入酒槽中与其他麦芽威士忌融合（此为混合纯麦威士忌），或者与其他麦芽威士忌及谷物威士忌融合（此为调和威士忌）。这里的谷物威士忌是由玉米等其他谷物酿造而成的蒸馏烈酒。

　　遗憾的是，许多威士忌生产商会在装瓶前用焦糖着色剂来给酒液上色。他们认为这样做的正当理由是，没有哪两个橡木桶里的陈酿变化是完全一样的，因此，来自同一品牌的威士忌，同样是陈酿十年，只要不出自同一个橡木桶，它们的颜色就不会一模一样。为了不让消费者感到困

肯塔基州洛雷托市的美格蒸馏厂中，波本橡木桶正在被烘烤。

肯塔基州洛雷托市的美格蒸馏厂中，威士忌正在橡木桶中陈酿。

惑，厂商便采用焦糖着色剂来确保每一瓶酒的颜色完全一致。不幸的是，添加着色剂也助长了一种广泛传播的错误观念，即威士忌颜色越深，代表其年份越高、质量越好。

　　许多威士忌生产商还会在装瓶之前对威士忌进行过滤。此举可以滤除少量被烧焦的橡木桶木屑，还可以去除脂肪酸和酯类，而这些物质会在消费者向威士忌里加水或冰块时让酒液显得混浊。评论家们则抱怨这种做法，因为他们认为这是以美观的名义牺牲了细微的风味。

威士忌酒瓶上标示的年份表明了威士忌在橡木桶中陈酿的时间。（威士忌在酒瓶中存放了多久则不会在酒标上标明。）一瓶标有"调和"或"混合"的威士忌里混有不同年份的威士忌酒液，而酒标上的年份必须按其中的最低酒龄来标示。还有一个不成文的经验之谈是，如果酒标上没有标明年份，那么这瓶威士忌的陈酿时间很有可能不到四年。

威士忌的种类

如果想让自己头疼，立竿见影的办法就是站在酒类专卖店或有售酒执照的商店的威士忌专区，然后开始阅读所有酒瓶上的酒标。那上面排列着一系列让人眼花缭乱的词汇，这些词还常常晦涩难懂，其目的就是将瓶中的威士忌同其他威士忌区别开来，比如"纯麦芽""酸麦芽""调和""小批量"等等。这些术语中，有些词是有用的，而其他词语的存在并不是为了让人看懂。

现在有许多威士忌词典，其中有对所有这些术语的定义，比如加文·D. 史密斯的《关于威士忌的 A 到 Z》。在这里我们只关注那些基本要素。一个人盯着装满威士忌酒瓶的架子，说到底想知道的就是"它们尝起来怎样"。只要对威士忌的基本类型熟悉了，你就能对一瓶酒的风味猜个大概。对于新手来说，有个简单直接的办法可以帮他们理清头绪，那就是按产酒国家对它们进行分类：美国威士忌常常带甜味；加拿大威士忌大都酒体①轻盈、果味浓郁；爱尔兰威士忌的风味一般比加拿大威士忌的更浓烈；苏格兰威士忌则大都带有烟熏味。

只要把上面的内容记熟了，你就可以进一步去接触更复杂的威士忌术语定义和下面列出的各种细分类型。这些词都是由法律和业内惯例所定的那些复杂定义的简化版本。

这些定义可以帮助我们探寻各种类型威士忌的风味，这一点非常重要。就像大麦汤喝起来肯定和玉米汤的味道不一样，爱尔兰威士忌（以大麦为原料）也不会和波本威士忌（以玉米为原料）有一样的特征。此外，与使用柱式蒸馏器制作出来的威士忌相比，使用壶式蒸馏器蒸馏的威士忌风味会更浓烈些（详见第四章）。

威士忌的种类

美国威士忌

波本威士忌：使用不少于51%的玉米为原料蒸馏而成的威士忌，并在经过烘烤的新橡木桶中陈酿。

玉米威士忌：使用不少于80%的玉米为原料蒸

---

① 酒体，专指酒液在舌面上的质量，是多种结构成分共同作用所产生的总体感受。酒精浓度和酸度是影响酒体感的主要因素，酒精浓度越高，酒体越重；酸度越高，酒体越轻。

馏而成的威士忌。

黑麦威士忌：使用不少于51%的黑麦为原料蒸馏而成的威士忌，并在经过烘烤的新橡木桶中陈酿。

田纳西威士忌：在田纳西州酿制的波本威士忌，通过木炭过滤之后，在经过烘烤的新橡木桶中陈酿。

加拿大威士忌

加拿大威士忌：由中性烈酒（几乎没有风味特征）混酿的威士忌，并在橡木桶中经过至少三年的陈酿。

爱尔兰威士忌

爱尔兰威士忌：使用大麦（有时会加入其他谷物）为原料蒸馏而成的威士忌，并在橡木桶中经过至少三年的陈酿。爱尔兰威士忌大多是由爱尔兰壶式蒸馏威士忌、麦芽威士忌和谷物威士忌调和的威士忌。

爱尔兰单一麦芽威士忌：由同一家蒸馏厂使用发芽大麦为原料，经过壶式蒸馏器蒸馏而成的威士忌。

爱尔兰壶式蒸馏威士忌：使用发芽和未发芽的大麦为原料，经过壶式蒸馏器蒸馏而成的威士忌。

爱尔兰谷物威士忌：使用大麦（有时会加入其他谷物）为原料，经过柱式蒸馏器蒸馏而成的威士忌。

苏格兰威士忌

苏格兰威士忌：使用大麦（有时会加入其他谷物）为原料蒸馏而成的威士忌，并在橡木桶中经过至少三年的陈酿。目前市场上销售的苏格兰威士忌大部分是调和威士忌。

单一麦芽苏格兰威士忌：由同一家蒸馏厂使用发芽大麦为原料，经过壶式蒸馏器蒸馏而成的威士忌。

单一谷物苏格兰威士忌：由同一家蒸馏厂使用发芽大麦（有时会加入其他谷物）为原料，经过柱式蒸馏器蒸馏而成的威士忌。

调和苏格兰威士忌：由一种或多种单一麦芽威士忌，以及一种或多种单一谷物威士忌混酿而成的威士忌。

显然，生产优质威士忌是个复杂的过程，需要大量的技巧和实践。一个人能弄清楚该怎么做，已经足以令人惊叹，更别说把它做好了。由此，我们也自然而然地想到这些问题——"谁发明了威士忌？什么时候发明的？"

［第二章］

早期历史

模糊与争议并存的威士忌起源

　　威士忌是什么时候发明的？由谁发明的？为什么发明的？这些直截了当的提问，答案五花八门，并没有一个定论。查阅不同的书会得到不同的答案，不过大体都会讲到，古希腊或近东的智者在至少 2000 年前发现了蒸馏的方法。这个技术通过某些途径传入了不列颠群岛，也许是圣帕特里克[①]和基督教传教士担任了这个传递者的角色，也可能是信仰伊斯兰教的摩尔人将这个方法带到了欧洲。后一种假设的支持者喜欢指出一个依据，即酒精的英文 alcohol 由阿拉伯词语 al-kohl 演变而来。人们还会在书里读到，这个方法传入不列颠群岛后，炼金术士或神职人员在做蒸馏实验时发现，通过蒸馏谷物竟可以得到美味的烈酒，最终这种酒被命名为威士忌。

　　和其他有价值的事物一样，威士忌也被一些人宣称是属于他们民族或祖国的宝贵遗产。有趣的是，其中有些人一边主张自己的声明，一边又承认他们没有任何的证据。《爱尔兰威士忌的 1000 年》的作者马拉奇·马吉就是个例证。这本书本可以是部迷人的作品，但书中的一个观点使它不再那么具有吸引力："没有人能确切地说出这一切是如何开始、何时开始的，但几乎可以肯定的是，威士忌，

---

① 圣帕特里克（约 385—461），出生于威尔士，少年时被绑架到爱尔兰成为奴隶，后来逃走，又冒着生命危险回到爱尔兰传播天主教，成为爱尔兰主教，后成为天主教圣人。

uisce beatha①，即生命之水，最早就诞生于爱尔兰。"马吉为什么能这么肯定呢？他自己也没说。马吉宣称，爱尔兰的修道士在去往中东的传教之行中学会了蒸馏技术，并在公元 500 多年或公元 600 多年将这项技术带回了爱尔兰。我们不清楚这个观点在"爱尔兰威士忌的 1000 年"这段故事中究竟起到了怎样的作用，但对于马吉来说毋庸置疑的是，苏格兰人是一群狡猾的人，"他们继而最终利用了大自然的慷慨馈赠"。

拉尔夫·斯特德曼是一位顽童般的威尔士艺术家，曾在其 1994 年的作品《酒瓶静物》中打趣这种"就是我们的发明"的态度：

> 戴克里先时代（205—305）②的埃及人是不折不扣的酒鬼。他们将消息告诉巴比伦人，巴比伦人再告诉希伯来人。一个昏昏欲睡的信使徒步将消息带到色雷斯，于是一个在当地旅游的凯尔特人开始叫卖皮带扣和皮革饰品……

斯特德曼宣称，在西欧，最早接触蒸馏技术的人是——你猜对了，就是威尔士人，他们在 1329 年通过蒸馏羊毛脂（一种由羊毛提炼出来的油脂）获得了强劲的烈酒。

任何事物的发明都可以归功于单独某一个人，这种想

---

① 爱尔兰语，用以指代威士忌，字面意思为"生命之水"。
② 此处原文疑有误，经查证应为 285—305。

法本身就是一个可疑的命题。没有一起研究，查尔斯·克罗和路易·迪科·迪奥龙都独自发明了彩色摄影；没有一起捣鼓，亚历山大·格拉汉姆·贝尔和伊莱沙·格雷也同时发明了电话。如果那些复杂的技术可以同时出现，为什么威士忌不可以呢？

我们说 whiskey，但他们说……

今天的我们要知道千万年前的人们做了什么，唯一的办法就是在千万年前的文字记录和手工制品中，找寻他们活动的迹象。举个例子，公元 600 年，发酵大麦汁煮沸后的蒸汽在高卢的一个洞穴壁上凝结成水珠，有个弓着身子的多毛野蛮人舔了舔，便发现了威士忌……如果没有这些记录和物品，我们怎么能知道这些过往？如果能将人类的整个存在历史都存进一个电脑硬盘里，那么历史学家的工作将变得简单得多。若任何人做过的任何事都被记录下来，并且能够全文搜索，那么只要你搜索相关的词汇，就能找到任何问题的答案了。

然而，即使真的有这样一个人类史实的宝库，历史学家在对威士忌的探索之路上仍然会受挫。这是因为在 1900 年以前，人们对用来指代致醉饮品的单词的含义几乎没有达成共识，语言表达上一片混乱。只看英语这门语言，比如 brandy（白兰地）这个词现在指的是用果汁蒸馏而成的烈酒，而在以前它可以指代任何酒，不论其原材料是水果、根茎还是谷物。同样的，liquor（烈酒）在以前常用来指代所有酒精饮料，包括啤酒和葡萄酒。

到目前为止，研究人员已经可以确定，whiskey（威士忌）一词最早是出现在1753年的一册《绅士杂志》中。该杂志报道称，都柏林的一家商店"售出了120加仑这种可恶的烈酒——威士忌(whiskey)"。根据《牛津英语词典》的注释，在此之前还有过这几种拼写形式：whiskee（也出现于1753年）、whisky（1746年）和whiskie（1715年）。

whiskey一词可能来源于usky（也拼作usquæ和husque），这个词显然是盖尔语词汇usque baugh的英语化，其发音是"oss-keh-baw"，意思是"生命之水"。然而usque baugh一词还有至少六种不同的拼写形式，包括usquebae（1715年）、uskebath（1713年）、usquebagh（1682年）、uscough baugh（1600年）、iskie bae（1583年）和uskebaeghe（1581年），这更是增加了威士忌名字来源的不确定性。usque baugh这个词已知的最早前身是uisce betha（1405年），显然，在这个词的发音上，爱尔兰人（发音为"oss-keh baw"）和苏格兰人（发音为"ooshkie-bayha"）明显不同。

现在，人们在阅读那些提到whiskey这个单词的早期印刷品时，往往很难理解这个词所指的究竟是哪种饮品。那些作者通常不会对这个词进行描述或定义，并且还会用它来称呼与我们所说的威士忌（以谷物为基础原料的酒）大不相同的东西。1600年，一名宫廷文职人员曾写到过

usquebaugh，指的是用葡萄干酿成的一种药酒。同样的，有一位作家曾在 1658 年提到一种由蜂蜜、葡萄酒和草药制成的饮品，并称之为 usquebach。1725 年，乔治·史密斯在他的《蒸馏全览》中写下了 usquebaugh 的配方，其中包含了发芽大麦、糖蜜、丁香、香菜、肉桂、坚果和糖。

因此，当我们在《克朗马克诺伊斯年鉴》（1405 年）中读到，蒙特利拉的理查德·马格内尔酋长因为喝了太多的 uisce betha 而逝世，我们并不能确定是哪种酒让他倒下的。

（极少）历史记录中的威士忌

那么历史记录中都记载了关于威士忌的什么内容呢？其实，在公元 1500 年之前，关于威士忌的历史记录很少，并且极其混乱。

发酵操作的历史可以追溯到大约 10,000 年前。公元前 8000 年，埃及和近东已经出现谷物的耕种和加工。根据考古证据，这些活动在欧洲的历史可以追溯到公元前 6000 年。

已有确凿的证据表明，古埃及人会将谷物酿成啤酒，并且还为此试验了各种各样的配方。古希伯来人也是如此。

欧洲人酿啤酒的知识是由中东传入还是由他们自己摸索出来的，目前尚不清楚。已有的证据表明，英国的啤酒酿造出现于公元前 3000 年，然而直到数世纪之后，人们才将发酵和蒸馏这两项技术结合在一起，威士忌的酿造这时也才开始出现。

有趣的是,《圣经》似乎有意将啤酒和葡萄酒与一种叫"烈酒"①的东西区分开来。《旧约·箴言篇》教导道:"葡萄酒能使人亵慢,烈酒使人喧嚷;凡因酒错误的,就无智慧"(20:1);"可以把烈酒给将亡的人喝,把葡萄酒给苦心的人喝"(31:6)。这种"烈酒"是指以谷物为基本原料酿造的烈酒吗?圣经并没有明确告诉我们。

古希腊人似乎已经对蒸馏的本质有了一个粗略的概念。亚里士多德在《气象学》(公元前 350 年)中就提到过他的观察:

> 当盐水变成水蒸气时,它的味道会变甜;水蒸气再次凝结时,并不会变回盐水。这是我通过实验得知的。这个现象在各种情况下都适用:葡萄酒和其他各种液体,经过蒸发和冷凝后会变成液态水。这样的水都受到了某种混合物的影响,也正是混合物的性质决定了它们的风味。

1970 年出版的《蒸馏艺术简史》是对现存史料最权威的综述回顾,作者 R. J. 福布斯在书中指出,亚历山大港的埃及化学家在公元 1 世纪或此后不久就掌握了这项技术。犹太女人玛利亚②等人曾在自己的作品中描述了他们这些早

---

① 原文为 strong drink。
② 犹太女人玛利亚是古希腊时期生活在埃及亚历山大港的一名女炼金术士(约公元前 3 世纪至公元前 2 世纪期间),发明了炼金术仪器及技术,被誉为西方炼金术的奠基人之一。

期科学家对不同液体和化学物质的操作——加热、研磨、混合和过滤。从示意图上可以看到，他们已掌握蒸馏所需的技术：一个用来将液体煮沸的球形容器（葫芦形蒸馏瓶），上面连接着一个狭窄且弯曲的瓶颈（蒸馏器），这是让蒸汽凝结成液体的地方；最后液体缓慢滴入一个容器（接收瓶）。值得注意的是，埃及人在公元前1500年就开始使用模具制作玻璃容器，到了公元1世纪，他们已经开始吹制玻璃容器。

此后，视蒸馏为一项转化和净化技术的阿拉伯人将蒸馏发展成了一个欣欣向荣的商业产业。阿拉伯的文献中也描述了通过蒸馏制作食品添加剂（如玫瑰香料）、精油、香水等市面商品所用的配方和技术。

在接下来的一千年里，关于蒸馏和其他早期炼金术操作的知识慢慢地向东北和西北方向传播。在传播过程中，

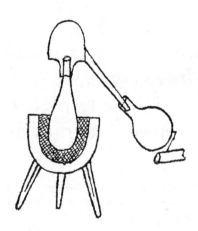

古代亚历山大港的葫芦
形蒸馏瓶和蒸馏器

这些技术得到了改进，并且还有了其他用途，如用于天然草药的生产。根据福布斯书中所述，可能是商人和其他旅行者将这些知识从亚历山大港带到北非，接着又传到西班牙，还从大马士革传到君士坦丁堡（现伊斯坦布尔），再到欧洲东部和中部。到了12世纪，欧洲人开始将阿拉伯的化学和科学论文翻译成拉丁文，例如英国的阿拉伯文化研究学者罗伯特·切斯特。1144年，他在西班牙待了一段时间，并在那里翻译了《炼金术全书》。回到英国之后，他应该也把他所了解到的知识分享给了更多的人。

　　没有人知道人类史上第一次尝试蒸馏经过发酵的液体是什么时候。亚里士多德曾记录过葡萄酒的蒸馏和冷凝；爱德华·吉本在他的《罗马帝国衰亡史》（1776—1789）中提到过两次"烈酒"，一次是讲5世纪时鞑靼人用马奶来酿酒，另一次是说匈奴村民用大麦酿造类似威士忌的烈酒，并称之为"卡慕斯"。然而直到15世纪，世界上才出现了有关生产蒸馏烈酒的强有力证据。身为哲学家兼教会人士的艾尔伯图斯·麦格努斯（1193—1280）在一篇论文中特别记录了一种白兰地的酿造配方，让现代人眼前一亮：

　　　　取一夸脱浓稠的陈年黑葡萄酒，加入生石灰（氧化钙）、硫黄粉、优质酒石（酒石酸氢钾）和白色食用盐，以上固体应全部研磨成细粉状，然后全部倒入带有蒸馏器、密封性良好的葫芦形蒸馏瓶；通过蒸馏，你就能得到"燃烧之水"①（烈性水）。这种液体需使用玻璃容器储存。

英国人罗杰·培根（1214?—1294?）与方济会修道士雷蒙德·勒尔（1234?—1315?）和鲁庇西萨的约翰（?—1366?）曾用文字记录下蒸馏烈酒在对抗生理衰退、延长寿命方面的能力，而佛罗伦萨医生撒迪厄斯·阿尔德罗蒂（1223—1295?）等人则设计出了从葡萄酒中提取纯酒精的改进方法。医生阿纳德斯·德·维拉·诺瓦（1235?—1311）显然也发现了蒸馏烈酒的力量："它会让某些人情绪过度爆发，也会让某些人充满活力、欣喜若狂且创意十足——这就是生命之水。"

同样的情况也出现在拉斐尔·霍林斯赫德 1577 年著的《英格兰、苏格兰和爱尔兰编年史》中：爱尔兰炼金术士、诗人和作家理查德·斯坦尼赫斯特（1547—1618）声称，威士忌是一种非凡的药物。"适度饮用，可延年益寿，可保持青春，可止咳化痰，可摆脱忧郁，可化解仇恨，可启迪心智，可振奋精神。"这还不够，斯坦尼赫斯特更深入地谈到，威士忌可以治疗浮肿，防止肾结石，将引发痉挛的肠道积气释出，缓解反胃症状，保持循环系统通畅和骨骼强健。"如果有规律地饮用，那确实是至高无上之酒。"

关于威士忌酿造的最早铁证到 1494 年才出现，当年的

---

① 原文 aqua ardens 为拉丁语，字面意思为"燃烧的水"，古时用来指可以点燃的蒸馏酒。

一个 16 世纪早期的欧洲蒸馏器

《苏格兰国库卷》中有这样一个条目记载："给修道士约翰·科尔 8 博耳①麦芽用于酿酒。"值得注意的是，这批麦芽的数量并不是个小数目，大概有 507 千克（1118 磅），可能可以酿造出大约 50 美制加仑②（190 升）酒，而最终会酿出多少还取决于酒精含量的高低。

一个 16 世纪晚期的欧洲蒸馏器

---

① 谷物度量单位，1 博耳等于 140 磅。
② 美国计量容积或体积的单位。1 美制加仑约等于 3.785 升。

此后，在《苏格兰国库卷》和《王室财政大臣账目》中，经常会提到被用来酿制蒸馏酒的谷物。显然，完全有可能在1494年之前就有人酿造威士忌了，可能是在爱尔兰、西班牙，甚至就像吉本声称的，也可能是在东欧偏远地区。然而，1494年关于修道士约翰的条目记载是目前已知的最早的参考文献，苏格兰的支持者们可能会为此感到自豪。

到了1506年，威士忌受到苏格兰知识分子的极大尊重，喜欢涉猎炼金术和医学领域的苏格兰国王詹姆斯四世甚至将威士忌生产垄断权授予了爱丁堡外科医师公会。政府的这一行为有两个方面值得关注。一方面，国王詹姆斯似乎将威士忌视为一种药物。当时的许多博学之士都认为，威士忌和各类蒸馏烈酒是药品而不是消遣饮品。将威士忌视为药物的观念也一直持续到20世纪（见第五章）。

另一方面，这项特权的授予实际上表明了政府有权决定谁能生产威士忌，谁不能生产威士忌。这样说来，爱丁堡公会里的知识分子便是慷慨解囊的施恩者，而城市里任何一个自酿威士忌的穷人都是理亏的。

我们所知的威士忌 vs. 过往的威士忌

威士忌生产商和公关公司经常夸耀自家的威士忌传承了古老酿法。他们声称，一个世纪或两个世纪或甚至三个世纪以来，他们都用同样的方式来酿造威士忌。他们的广告描绘了这样一幅酿制威士忌的画面：一位遵循传统的老

一个来自宾夕法尼亚州兰开斯特市的 18 世纪的蒸馏器

人，一双结实粗糙的手，一张永远不变的秘密配方。不久前我参加了一场威士忌品鉴会，看到一位知名苏格兰威士忌公司的代表面不改色地宣称，他们公司现在生产的威士忌尝起来和 150 年前的一模一样。

这类说法大都是无稽之谈，今时今日的威士忌和 20 世纪以前的威士忌早已截然不同。在过去的 125 年里，各国政府纷纷制定法律法规，定义了威士忌的基本属性，并为酿酒可能用到的原料、必须使用的橡木桶类型以及酒的陈酿时间设定了标准。然而在此之前，在威士忌广告中备受推崇的往昔，几乎没有规定被执行，消费者也不清楚还有什么东西会被扔进麦芽浆里，可能是土豆、食糖、燕麦、芜青或其他任何酿酒师能搞到手的东西。通常情况下，那时候的酒也不会经过橡木桶陈酿，就算有陈酿过程，威士忌生产商也是使用任何他们能找到的橡木桶，而不会考虑其内部是不是全新的、烘烤过的，是不是沾有雪莉酒或葡萄酒的气味，甚至是腌鱼的臭味。而且，因为加入了蜂蜜和草本植物（如百里香、茴芹或薄荷）来调味，这些所谓的"威士忌"常常和杜松子酒或利口酒几乎没有什么区别。

人们对威士忌的早期历史知之甚少，很大程度上和当时威士忌的生产性质有关——它并不是面向市场供大众消费的商品。现有的证据表明，威士忌曾是由修道士和炼金术士生产的，并且最初被认为是一种药品。农民酿制威士

18世纪用火加热的蒸锅式蒸馏器复原图，这种设备与16世纪末期的欧洲蒸馏器非常相似。

忌也是完全有可能的，他们居住在城市之外，纯朴且大多目不识丁。因此，如果他们确实自己生产威士忌，也不会费心去记录自己的成就，这一点并不让人意外。他们为什么需要将酿酒事宜以书面形式记下来呢？

综上所述，直到政府开始介入威士忌生产行业，才出现了关于威士忌生产的完善文件和书面记录。在接下来的章节中我们会看到，詹姆斯四世对威士忌生产垄断权的授予，激起了个体与政府争夺威士忌生产自由权的长期斗争的千层浪。

苏格兰威士忌

总的来说，苏格兰威士忌的发展历程算是顺风顺水。苏格兰人似乎是最早生产威士忌的，但其产量很快就被爱尔兰人超过。然而到了 19 世纪中期，苏格兰威士忌又逐渐崛起。尽管遭遇了禁酒运动、经济萧条和世界大战，风味多样、种类丰富的苏格兰威士忌还是赢得了世界各国人民的喜爱。

早期历史

　　本书第二章提到，苏格兰人于 15 世纪 90 年代就开始蒸馏烈酒，此后不久，国王詹姆斯四世便将威士忌的生产垄断权授予爱丁堡。到了 1579 年，将淀粉植物蒸馏成酒的这种技术已经广泛普及。因为生产威士忌需要消耗大量粮食，苏格兰政府担心由此会导致饥荒，所以当局限制了威士忌的生产，除了伯爵、勋爵、男爵和绅士，其他人一律禁止酿造威士忌，而上述这些拥有特权的人，也只能自产自用，不得售卖。然而到了 1644 年，随着威士忌生产技术的传播，苏格兰议会看到了威士忌所能带来的财富，便开始对其征税。

　　威士忌生产技术究竟是由城市中心的知识分子传播到乡村，还是反过来从乡村传入城市，我们不得而知。但无论事实如何，在爱丁堡和格拉斯哥这两个苏格兰文明的中心，以及北部（奥克尼群岛）和西部（艾雷岛）岛屿上偏远的农田和村庄中，都曾有蒸馏器工作的身影。马丁·马丁在他的作品《苏格兰西部岛屿描述》（1695 年）中记录了让他大吃一惊的场景：

> 他们有大量的谷物，于是当地人酿制出了好几种类型的烈酒，比如常见的 Usquebaugh[①]。另一种烈酒叫 Trestarig[②]，即 Acqu-vitae[③]，这种酒需经三次蒸馏，强劲且辛辣；第三种是经过四次蒸馏的烈酒，被当地人叫作 Usquebaugh-baul[④]……喝第一口时，就会影响到全身各个部位：这种"最后之酒"喝两勺就足够了；一旦超过这个量，就会立刻让人停止呼吸，危及生命。

威士忌和英国的统治

苏格兰威士忌的发展能大致上顺风顺水，得益于许多因素——苏格兰人的节俭、勇敢，当地地形地貌（肥沃的土壤，大量的泥煤，许多优质水源，等等），以及苏格兰早期对资本主义的接纳。

还有一个被低估的因素是苏格兰和英格兰之间的关系。尽管耗时两个世纪，苏格兰人和英格兰人最终还是达成了

---

① 盖尔语，意为"生命之水"。
② 盖尔语，包含"三次"和"中东亚力酒"（一种起源于阿拉伯的烈酒）之意。
③ 拉丁语，意为"生命之水"。
④ 源自盖尔语，意为"一种对生命有害的水"。

政治和经济上的合作。这让苏格兰人得以发展威士忌产业，并在接下来几个世纪里将它不断壮大。（在关于爱尔兰威士忌的第四章里将讲述与此相反的案例。）

早在 1603 年两国开始由同一位国王统治之时，双方就开始了认真的合作。到了 1707 年，议会联盟将苏格兰议会并入设立于威斯敏斯特的代表机构，两国经济正式交织在了一起。协议中还提到，两国生产的蒸馏烈酒的税率是均等的。

当然，苏格兰和英格兰之间的政治合作不应被过分夸大——敌意仍然存在。苏格兰低地比（北部的）高地更适应王权的统治；前者的居民是比较温文尔雅的新教徒，而后者的居民多为天主教徒，在农村地区过着以宗族为纽带的生活。正如一段无名打油诗所说：

> 哦，特威德①的北边有什么？
>
> 怪兽，和弯着膝盖的多毛山人！
>
> 还有唤醒亡者的音乐！
>
> 去那里探险真是危险啊！
>
> 可怕的哈吉斯②出没于雪地，
>
> 马形水鬼③等着灾祸将你啃噬，

---

① 特威德河，历史上曾为苏格兰和英格兰的分界线。
② 一道传统苏格兰菜，也称肉馅羊肚或羊肚杂碎布丁。做法是先将羊的胃掏空，里面塞进剁碎的心、肝、肺等羊内脏，再加上燕麦、洋葱、羊油、盐、香辣调味料和高汤等，制成袋（现在常用香肠衣来代替羊胃），水煮约三小时，到鼓胀而成。
③ 苏格兰民间传说，恶灵化成马的形象，诱人自溺或预告人们将溺死。

除了生燕麦这里没有东西可吃，

但我仍被告知那里有威士忌啊！

英格兰人和苏格兰人之间发生过多次冲突。始于1745年的那个事件大概是其中最出名的：当时被称为"英俊王子查理"的查尔斯·爱德华·斯图亚特（1720—1788）领导了一场高地起义，希望为家族夺回王位。战争的结局是残酷的——战争第二年，他麾下许多身穿苏格兰短裙的士兵在卡洛登被屠杀，威斯敏斯特政府随后禁止苏格兰人持有武器和穿着苏格兰短裙及"高地服装"（这些禁令于1782年被废除）。

威士忌税是两国之间的一个痛处。1643年，议会首次对蒸馏烈酒进行征税，并且每当英格兰发觉自身处于战争状态时，他们就会盯上威士忌产业的财富。用于酿造苏格兰威士忌的麦芽、蒸馏器、从蒸馏器中流淌出的烈酒等物，都成了征税对象。1781年，议会禁止了私人蒸馏，且税务

18世纪后期，四个小商贩出售"木棍、民谣、贝德福德郡动物尾巴以及如樱桃酒般美味的苏格兰威士忌"。

当局获得授权，有权没收蒸馏器等用于威士忌生产或运输的任何物品，包括马匹和运货马车。

值得赞扬的是，伦敦政府确实从中吸取了教训。1816年，议会通过了《小型蒸馏器法案》，以降低威士忌关税。在接下来的几十年里，其修正案进一步降低了合法生产的威士忌的关税，并对非法生产和非法消费苏格兰威士忌的行为加大了处罚力度。

但是政府并不完全信任酿酒商。政府要求，每一家拥有执照的蒸馏厂都必须为一名常驻税务人员提供住处，这名税务人员将决定蒸馏厂需要上缴的税额。通过测量倒进蒸馏器的酒醪量和最终产出的烈酒量，税务人员会向蒸馏厂开出相应的税金账单。议会还下令使用烈酒保险箱，这是

布鲁克莱迪蒸馏厂的烈酒保险箱，位于苏格兰艾雷岛。

一种由玻璃和黄铜制成的密封箱，可以防止威士忌生产商在税务人员测量之前将威士忌从蒸馏器中转移以逃避税收。

躲避税务人员

在《国民财富的性质和原因的研究》[①]（1776年）一书中，亚当·斯密清楚地看到了政府政策中存在的许多问题：他们"将某事认定为罪行，即使这件事的本质绝非如此"。许多人觉得，选择将谷物制成烈酒是件私人的事情。人们开始说威士忌是"无辜的"，这并非无缘无故。不仅如此，许多政策的制定也没有将刺激因素考虑进来。正如斯密指出的那样，对酿酒商征的税越多，对方就越有非法蒸馏的动机。

实际上，当局似乎一直相信他们可以控制蒸馏的行为，只需通过立法告诉人们不能这样做，或者在人们这样做的时候课以重税就行了。一开始人们还是服从管理的，而且当苏格兰人对法律嗤之以鼻时，政府武装代表和令人畏惧的税收官或税务人员便会被派上阵。反常的是，旨在增加税收的法律往往会让税收减少，因为合法的威士忌产量骤降，而非法的威士忌产量激增。仅在爱丁堡就有大概400台蒸馏器处于生产状态。为了不让税务人员发现，蒸馏器会被巧妙地隐藏起来——天桥下，房屋地板底下（蒸汽和

---

① 简称《国富论》。

烟通过烟囱排出），甚至还会被藏在镇上的钟楼里。橡木桶和大罐子被埋在院子里，被藏在树上，还被装进棺材里偷运。据传，过去格洛斯特郡奥尔德伯里城的农民为了躲避政府人员，常常在他们装有非法威士忌的橡木桶外标上"洗羊药水"（一种有毒的化学制品，用于消除羊身上的虫子和真菌）的字样。因此，即使在今天，人们还是能在世界上最好的一些威士忌商店里，看到被商家顽皮地贴上"洗羊药水"标签的威士忌。

有件事相当愚蠢，就是当局试图通过给上交蒸馏设备者提供现金奖励的方法减少非法生产。有些威士忌生产商从这项政策中寻得了好处。他们将自家旧得不能再用的蒸馏器上交，再用奖励的现金购入材料，制作新的蒸馏器。

谈论非法威士忌这个话题时，就不能不提到那些将其带到市场上的人——走私者。虽然许多写作者对走私威士忌轻描淡写地一笔带过，但事实上，这往往是一种极不体面的生活方式。当时一位名叫伊恩·麦克唐纳的税务人员记录了 19 世纪苏格兰高地的走私情况："我辖区内的大多数走私者，我私下都认识。除了极少数的个例，他们都是最穷苦的民众。"这些走私者的房子处于失修状态，他们的田地也很少被照料，因为他们夜间工作劳累，白天都在睡觉。麦克唐纳指出，这种生活方式就像喝了大量威士忌一样，让他们筋疲力尽。他在《高地走私》（1914 年）中写道："渐渐地，他们的男子气概被削弱，他们的廉耻心变得麻木，他们会变成暴力的违法者和无耻的骗子。"走私者经

1914 年左右，苏格兰威士忌蒸馏器和供人栖身的棚屋。

常袭击甚至杀害做着本职工作的税务人员——这并非什么英勇行为。

虽然蒸馏师和税务人员之间经常发生争执，不过有时他们的意见也能达成一致。当税务人员从蒸馏器里取出的酒超过测量所需时，蒸馏师可能会睁一只眼闭一只眼。作为回报，税务人员也会在记录烈酒产量的时候减少数字，以降低其纳税额。

当然，蒸馏师和税务人员也能让彼此的生活变得异常艰难。税务人员可以在官方报告中写下针对蒸馏师的坏话；作为回应，蒸馏师可能会折磨税务人员。由于每次蒸馏开始时税务人员都必须在场，所以蒸馏师便可以在凌晨三点启动蒸馏器，把税务人员从床上赶下来。

不过随着时间的推移，政府重新考虑了税收和执法的方式，当局和威士忌生产商之间的关系也得到了改善。新政策采取激励措施，鼓励威士忌生产商合法经营，而不是非法生产。激励措施的改变，使合法生产威士忌远比非法生产更具吸引力。此外，政府还通过制定良好的生产规范，比如使用橡木桶陈酿威士忌，来帮助苏格兰生产商集体提高产品质量。这是一种双赢的关系：蒸馏厂将合法品牌推向市场，政府取得相应的税收，酒客则可以以合适的价格轻松购买到优质的苏格兰威士忌。

因此，税务代表和走私者之间的武装冲突减少了，没收非法蒸馏器的情况也减少了。这种富有成效的工作关系得到迅速发展。1983 年时，税务人员已不再徘徊于蒸馏厂附近。测算产量并向税务部门报告的职责也被转交到了蒸馏厂经理手中。如今，烈酒保险箱已成为昔日的产物，法律不再规定强制使用。简而言之，随着违法行为的减少和信任的建立，以前那种警察和疑犯的关系，现在变成了外部审计员和生产者的关系。

不断壮大的苏格兰威士忌

随着 19 世纪的发展，苏格兰和英格兰加强了彼此之间的合作关系。英国的一些上流社会人士会定期到苏格兰度假。最著名的就是维多利亚女王和阿尔伯特亲王于 1848 年在高地巴尔莫勒尔城堡开始的夏日避暑之行。在那里，他们玩着入乡随俗的游戏，有时穿着格纹呢绒衫，还会从附

入乡随俗的高地风格：威尔士亲王和阿伯丁侯爵，以及在后方的约克公爵和亨利亲王。照片摄于1920年左右。

近的蓝勋蒸馏厂订购几桶威士忌——后来女王向这家蒸馏厂的负责人约翰·贝格授予了皇家委任认证，这家蒸馏厂便更名为"皇家蓝勋"。（女王没有在爱尔兰建立类似的住所。）

随着大英帝国在全球范围内的扩张，苏格兰威士忌也随之扩大国际市场版图，让许多人变成了它的信徒。一艘艘满载威士忌（尤其是调和麦芽威士忌）的货船远航至巴哈马、埃及、印度、澳大利亚、新西兰和南美洲，威士忌产量也随之激增。迅速扩张和工业化的美国，尽管自身已拥有强大的威士忌产业，但本国酒客也喜欢上了略带烟熏风味的威士忌。

虽然如此，但对于苏格兰威士忌产业来说，20世纪上半叶还是残酷的。劳合·乔治①，第一次世界大战，经济大萧条，第二次世界大战，以及随之而来的定量配给，这些都导致了许多威士忌生产商的破产。最黑暗的时期要数1943 年，当时苏格兰威士忌几乎全部停产。

尽管如此，苏格兰的蒸馏厂还是存活了下来。产品的多样化起到了帮助作用——通过使用连续式蒸馏器，苏格兰的蒸馏师们在两次世界大战中为英国生产了数千万加仑的工业酒精。（在此期间，爱尔兰和英格兰之间于 20 世纪30 年代爆发了报复性的贸易战，而且爱尔兰在第二次世界大战中保持中立。）此外，苏格兰人还受益于先前签订的海外协议。当轴心国战败，世界开始恢复正常时，苏格兰生产商立刻开始向新老客户推销他们生产的产品。日本代表着一个重要的新兴市场。到 20 世纪 70 年代中期，每年有超过 700 万美制加仑（2650 万升）的苏格兰威士忌涌入日

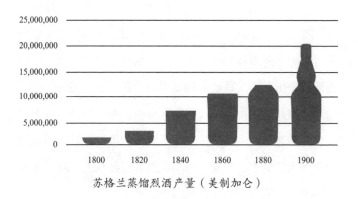

苏格兰蒸馏烈酒产量（美制加仑）

---

① 英国政治家，曾任英国首相。参与推行禁酒运动；进行酒牌法改革；主张加重对奢侈品、酒精、烟草等的课税。

本，这个数据着实令人震惊。

因此，在 1875 年到 1975 年间，爱尔兰威士忌产业崩溃，蒸馏厂数量从 65 家暴跌到只剩 3 家，而与此同时，苏格兰威士忌蒸馏厂的数量却从 112 家增长到 122 家。

最终，为了跟上全球市场不断增长的需求，苏格兰的

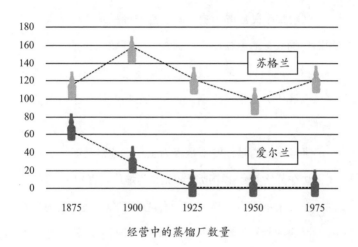

经营中的蒸馏厂数量

苏格兰威士忌出口量（美制加仑）

蒸馏厂不断地扩大生产规模。20世纪下半叶，苏格兰威士忌出口量猛增，从1950年的660万美制加仑（2500万升）增长到2000年的近6800万（2.57亿升）。

苏格兰威士忌和"苏格兰性"

纵观整个民族的历史，威士忌曾经只是苏格兰人生产和消费的众多酒精饮品中的一种。最开始应该是啤酒，然后是白兰地和利口酒，最后才是威士忌。然而，苏格兰威士忌已经成为苏格兰之酒，也是"苏格兰性"不可或缺的一部分。

值得一提的是，要不是因为虫子，苏格兰威士忌可能还无法在苏格兰文化中取得如此高的地位。直到19世纪末期，在后来组成当今英国的所有地区，白兰地（由葡萄酒蒸馏而成的烈酒）都一直深受消费者的喜爱，不论是富人还是穷人。后来，大自然母亲插手打破了这个局面——葡萄根瘤蚜，一种吸食汁液的害虫，于19世纪60年代开始破坏法国的葡萄园。葡萄酒产量骤降，白兰地变得异常昂贵，而且很难买到。然而彼时的威士忌价格便宜，产量充足。

起初，威士忌和文化之间的这种融合，是因为反抗议会早期针对威士忌的政策而产生的。当政府对威士忌征税，并将无证生产定为犯罪行为时，许多苏格兰人仍然坚持己见。原本只是一种饮品的威士忌，这时变成了一项事业。

罗伯特·彭斯（1759—1796）是苏格兰最著名的早期威

士忌宣传家和浪漫主义者。

> 让其他诗人去引起纷争，
>
> 关于葡萄藤、葡萄酒和喝醉的酒神……
>
> 我唱颂这苏格兰之酒，正是她成就了我们，
>
> 不论在杯中，还是罐里。
>
> 噢，我的缪斯女神！这美好的苏格兰老酒……
>
> 赋予我灵感，哪怕我说不清话，眨着眼，
>
> 依旧唱颂他们的名字！

彭斯把威士忌政治化，将其与苏格兰身份认同感和民族主义融合在一起。"苏格兰，我敬爱的老母亲！……自由与威士忌同在，举杯吧朋友！"

作家埃涅阿斯·麦克唐纳（又名乔治·马尔科姆·汤姆森，1899—1996）在他 1930 年的经典之作《威士忌》一书中，也表达了类似的关于苏格兰威士忌的浪漫观点：

> 威士忌的英雄时代之序曲就在那时揭开，当它在山上被悬赏猎捕，犹如斯图亚特王子；当忠诚无畏的人们用自家简陋的小木屋为它提供庇护；当关于它的学问被人秘密流传，恰似饱受迫害和禁锢的宗教所遵循的信条。

苏格兰威士忌是"苏格兰性"的一部分——这一观念

被威士忌产业的营销进一步加深。1853年英国政府停止征收广告税，并于1860年批准以瓶装形式销售烈酒。苏格兰人把握利好政策，开始满怀热情地推广威士忌。他们将威士忌描绘成绅士的饮品，一种展现个人品位和地位的消费品。苏格兰威士忌的推广宣传中也对苏格兰生活进行了近乎漫画般的理想化描绘——苏格兰人穿着苏格兰短裙，猎杀牡鹿、打高尔夫、飞蝇钓，还随身携带着风笛。1883年，汤米·德华[①]开始用一位身着苏格兰短裙和高地服饰的风笛手来做广告宣传自家的调和威士忌，这个做法延续至今。麦格雷戈家族、金鹰堡和酒瓶上贴着格子呢标签的麦克杜康等品牌强化了这一形象。最早的一批电影广告中，有一个就展现了四个身穿苏格兰短裙的男人在帝王威士忌的横幅前跳舞的场景。

当然，并非所有广告都是这种类型，有些就将苏格兰威士忌吹捧为一种健康补品。苏格兰威士忌品牌坎布斯称自家的酒是一种"有益健康的兴奋剂"，说它"有助于健康，且对脑部和肝脏均无危害"。其他的一些宣传活动则突然彻底走上了纯粹荒唐滑稽的道路。帕蒂森有限公司就曾带着鹦鹉去到各个酒吧，让鹦鹉尖声叫着"喝帕蒂森威士忌！"

随着20世纪的发展，苏格兰人的"苏格兰性"形象发生了些许变化——真诚的田园主义少了，机智和欢乐多了。"狡猾的苏格兰人"总有办法搞到自己要喝的威士忌，有些

---

① 帝王威士忌品牌创始人约翰·德华的儿子。

20世纪10年代滑稽的宣传广告：苏格兰人喝着桑迪·麦克唐纳苏格兰威士忌。

广告宣传便利用了这一旧形象。1949年的滑稽片《荒岛酒池》中就有类似的场景：一艘船在苏格兰海岸失事，附近居民与海关当局斗智斗勇，试图抢夺船上装载的5万箱苏格兰威士忌。《荒岛酒池》并非编剧的虚构，而是根据真实故事改编。

1820 NOT OUT

20世纪00年代，一则尊尼获加威士忌的广告刻画了一个板球比赛中"迈步向前的绅士"的形象。

　　与此同时，执杯者协会①在为苏格兰威士忌与苏格兰性的结合增添一丝尊贵之气的同时，近年来也一直试图延续

---

① 苏格兰威士忌产业中为国际公认的行业协会，旨在表彰世界各地对苏格兰威士忌有杰出贡献的人士。

这两者之间的联系。这一团体宣扬赞颂苏格兰威士忌的高贵出身和社会名望。（执杯者协会中的"杯"指的是苏格兰双耳小酒杯，外形有点像一只带两个把手的碗。）威士忌产业从业人员于1988年创立了这个组织，不过其风格显然是老派的。该组织拥有自己的纹章，其口号是"Uisgebeatha Gu Brath[①]"，即"生命之水永存"。他们会在布莱尔城堡举办宴会（这座城堡自然位于苏格兰高地），人们穿着苏格兰短裙，吹奏着风笛，叉起羊肚杂碎布丁，大口畅饮苏格兰威士忌。

物极必反

当然，关于苏格兰人在生活中的处境也有一些不太愉快的言论。托马斯·克罗斯兰在其臭名昭著的《难以言说的苏格兰人》（1902年）一书中发表了一通批判，就文化自卑、道德败坏及整体"平庸"这几个方面对苏格兰进行贬低和抨击。克罗斯兰，一个被同时代人称为"歇斯底里反苏格兰"的英国人，对苏格兰所谓的高犯罪率怒声斥责，声称这个民族饱受嗜酒成性之苦。"苏格兰已经成为世界上头号酒鬼国家。"克罗斯兰愤然说道。

早餐是威士忌，午餐是威士忌，晚餐是威士忌；

---

① 盖尔语。

> 与朋友见面时是威士忌，任何商务会议上都是威士
> 忌；去教堂之前是威士忌，出来时是威士忌……在
> 你健康时是威士忌，在你生病时是威士忌，几乎在
> 你一出生时就是威士忌，在你离世前的最后一件事
> 是威士忌——这就是苏格兰。

在克罗斯兰的描述下，典型苏格兰人的状态是大起大落的，一会儿是被威士忌刺激起来的兴奋，一会儿又变成"死气沉沉"的清醒。"你跟他说话，得到的回应只有咕哝声；他没有笑容……他郁郁寡欢，言语粗鲁，头脑迟钝。"

后来，一些没那么暴躁的作家，如乔治·道格拉斯·布朗、约翰·麦克杜格尔·海等，也在书中描述了苏格兰威士忌消费中丑陋的阴暗面——酗酒导致的犯罪、残酷暴力和个人毁灭。

这些描述并非只是作者的虚构；不论何地，只要有大量的廉价蒸馏烈酒出现，就会有严重到令人不安的酗酒现象出现。就像大量价贱如土的杜松子酒涌入城市地区，使英格兰深受其苦一样，苏格兰的部分地区也同样没能抵挡住威士忌。

在 19 世纪 30 年代，已满 15 岁的苏格兰人除了喝非法威士忌和其他酒精饮品之外，每周还要喝大约一品脱的合法威士忌。在这个国家的部分地区，威士忌完全融入了苏格兰人的日常生活。无论在什么场合，人们都会举起一小杯威士忌。一场婚礼？喝！有婴儿出生？喝！有人去世？喝！

Nae Macallan... Nae Fish.

A STORY IS TOLD of Donald, a revered ghillie in years gone by on a certain loch of our acquaintance.

It was a bad morning for trout, the water a glassy calm.

Donald toiled all morning at the oars while his cargo of two London businessmen caught nothing.

As lunchtime neared, Donald began to look forward to the lustrous sherry-gold depths of the bottle of The Macallan Malt Whisky which was the customary reward for a deserving ghillie.

But the otiose Sassenachs had other ideas.

"No fish, Donald", they cried. "Then no whisky."

Donald said nothing and ate his lunch at some remove. But the iron had entered his soul.

The wind rose. And all afternoon while every other boat on the loch was landing an almost miraculous draught of trout, Donald rowed his clients slowly up and down the one unruffled stretch of water.

When evening came, he deposited his fishless clients on the bank and surveyed them gravely as they rifled through their treasuries of insult, goggling like the trout they had so signally failed to capture.

"Nae Macallan" said Donald, at last "Nae fish." And rowed off into the gloaming.

The MACALLAN. The MALT.

20世纪80年代一则幽默的麦卡伦威士忌广告。图中文字大意是：唐纳德是一个备受夸赞的苏格兰捕鱼向导。某天，他载了两个英格兰商人上船捕鱼，由于上午的天气不适合捕鱼，虽然他拼尽全力帮忙，但两个英格兰人还是一无所获。到了中午，按照当地的习俗，两个客人本应将美味的麦卡伦单一麦芽威士忌作为酬劳分给唐纳德，但他们却说道："没有鱼，就没有威士忌。"下午刮起了大风，湖面波涛翻滚，非常适合捕鳟鱼。就在其他渔船都在激浪中猎鱼时，唐纳德却带着两个英格兰人慢悠悠地划着船，来回于一片风平浪静的水域。到了傍晚，唐纳德将两个仍然一无所获的英格兰人送上了岸，并说道："没有麦卡伦，就没有鱼。"之后便在暮色中划船离开了。

不同地区饮用威士忌的大体模式差异显著。苏格兰的农村地区很少有酒吧。比如在设得兰群岛，平均每1000人才有一家酒吧。于是，农村人从早到晚都在家里喝威士忌，每次喝的量不多，破晓时喝一点，工作间隙休息时喝一点，黄昏时再喝一点。

人口密集的城市地区则是另一回事了。比如在格拉斯哥，平均每130人就有一家酒吧；这还没把非法经营的地下酒吧算进来，在这些店里，极其强劲的廉价威士忌涌流不息，通常又被叫作"致命一酒"。糟糕的城市环境和大男子主义都助长了酗酒之风。商人们在店里的规矩可以说是彻底的巴洛克式[①]。他们设立了饮酒基金，并要求员工缴纳一小部分钱，无论其表现好坏。你照看的火熄灭了？把钱交到饮酒基金！你得到加薪了吗？把钱交到饮酒基金！只要罐里的钱足够多，员工们就会狂饮作乐，一夜之间把钱都挥霍在买威士忌上。渔业城镇和工业城镇尤为夸张，就像被威士忌浸透了一样。乔治·贝尔——世纪中叶的一位反酒精倡导者，对他在贫民窟里看到的一切表示谴责。"从没有牙齿的婴幼儿，到没有牙齿的老人，所有居民……都喝威士忌。周六晚上和周日早上上演的醉酒戏码，难以用笔墨描述。那样的场景太可怕了。"

让人高兴的是，虽然19世纪苏格兰威士忌的产量猛增，但苏格兰人的威士忌消费量却随着时间的推移显著下降。

---

① 欧洲17世纪时的一种艺术风格，这种风格在反对僵化的古典形式、追求自由奔放的格调和表达世俗情趣等方面起了重要作用。

苏格兰艾雷岛的雅柏蒸馏厂

1900 年，苏格兰成年人人均每周威士忌消耗量勉强超过半品脱。到了 1940 年，人均消耗量降到了每周几小口。

为什么下降？这很难说。有人认为，体育和娱乐的发展让人们在空闲时间有更多的选择，而不单单是喝酒。19 世纪前期，一些禁酒协会突然开始出现，它们至少也能让一部分人不再酗酒。毫无疑问，人们的生活方式普遍发生了转变。一个人如果花一整晚的时间喝得酩酊大醉并把事情搞得一团糟，则会被视为行为失控的下层阶级。（爱尔兰和美国也出现了这样的转变。）

此外，政府行为也影响颇大。1853 年的《福布斯－麦肯基法案》大大缩短了酒吧营业时间，而 1855 年的《甲基化酒精法案》给政府增加了打击地下酒吧和非法饮酒俱乐部的权力。这是个缓慢但稳步生效的过程，政府的一系列举措缩短了每天可购买威士忌的时间，并减少了其销售场所的数量。这些政策同几十年来逐步提高的税率一道，增加了个人购买威士忌的金钱、时间和精力成本。

地　形

苏格兰在威士忌世界中是独一无二的，其多样的地形，长期以来造就了各式各样的威士忌，尤其是那些瓶装单一麦芽威士忌。（调和威士忌本质上缺少了这些地域差异。）

所有的苏格兰威士忌都是用水、经泥煤烘干的大麦和酵母酿造而成的。然而，一个从未品尝过苏格兰威士忌的人，也能轻易分辨出苏格兰高地威士忌（如格兰威特 12 年）和艾雷岛威士忌（如拉弗格 10 年）之间的区别。前者风味清淡，略带烟熏味，有鲜花和蜂蜜的香气；后者则让口腔和鼻腔都充满了烟熏味、碘味和海草风味。

苏格兰威士忌的多样性是多种因素共同作用的结果，包括不同的水源、不同的大麦品种、不同的酵母株系、不同的威士忌配方（酿酒谷物配比）、不同的蒸馏设备，以及桶陈地点不同的气候。其中一些因素确实取决于当地地形，不过，剩下的就取决于威士忌生产商的决定了。尽管如此，苏格兰威士忌产业还是将单一麦芽威士忌划分为五大产区。

艾雷岛产的威士忌可能是风味最浓郁的麦芽威士忌，闻起来和尝起来通常都有海水、碘和烟熏的味道。最知名的艾雷岛威士忌有雅柏、波摩、拉弗格和乐加维林。

相反，那些产自邓迪市和格里诺克市南部的苏格兰低地威士忌往往风味柔和，而且很少带有艾雷岛的那些典型风味。现在只有几家低地蒸馏厂还在继续生产——欧肯特轩、磐火和格兰昆奇。

艾雷岛的东边就是坎贝尔敦，这里曾经是威士忌生产的一大基地。坎贝尔敦以前有30家蒸馏厂，而现在只有格兰帝和备受尊崇的云顶蒸馏厂还在经营中。坎贝尔敦生产的威士忌与艾雷岛的风格相似，不过风味通常没有那么强烈，而且还带有其他比较清淡的香气。

一些最有名的苏格兰威士忌，如格兰威特、格兰杰和泰斯卡，都产自苏格兰高地，这片地区就位于苏格兰低地北边并与之接壤。该产区的蒸馏厂生产的威士忌种类多得惊人。也许高地威士忌唯一的共同特征，就是风味都偏果味且整体品质优异。

苏格兰的蒸馏厂大概有一半都位于高地的东北部，该地区被叫作斯佩塞。该产区有一些狂热的支持者，他们宣称斯佩塞是苏格兰威士忌的皇冠之珠。毫无疑问，斯佩塞出产了包括亚伯乐、百富和格兰路思在内的许多顶级威士忌。然而，要说出斯佩塞威士忌和高地威士忌之间有什么特征差异是很难的。实际上，从风味清淡、带有青草香气的格兰菲迪，到风味浓郁、果香扑鼻的麦卡伦，斯佩塞威士忌本身的特征就非常多样化。

20 世纪末的苏格兰威士忌

　　苏格兰威士忌在过去的一个世纪中战绩辉煌。产量直线上升，在世纪末超过了 1.9 亿加仑，可供选择的酒款数量也成倍增加。从低价的白马和威雀（每升约 11 英镑或 18 美元），到高端的芝华士 18 年（50 英镑或 50 美元）和超级昂贵的尊尼获加蓝牌英皇乔治五世（535 英镑或 400 美元），市面上提供的高品质苏格兰调和威士忌多到令人眼花缭乱。现在，调和威士忌在出产时的酒龄也越来越长了。以廉价酒款（25 英镑或 15 美元）闻名的顺风，以往生产的威士忌只陈酿几年，现在也发售 12 年陈、15 年陈、18 年陈和 25 年陈的威士忌了。

　　历经几十年的黑暗时期之后，桶陈苏格兰威士忌再次

艾雷岛格兰杰蒸馏厂中用来陈酿威士忌的橡木桶

爱尔兰威士忌

爱尔兰威士忌的历史是一个动荡的故事。它来历不明、出身低下，在 19 世纪发展到巅峰，却又遭遇了几乎宣告其终结的重挫。然而在过去的半个世纪里，爱尔兰威士忌经历了一场拉撒路式①的复苏。虽然现有的蒸馏厂及品牌比以前少了很多，但是今天的爱尔兰威士忌品质一流，受到了全世界的认可。

早期历史

烈酒蒸馏技术在爱尔兰存在的痕迹可以追溯到 12 世纪。1170 年，英格兰国王亨利二世派兵进军爱尔兰，目的是使第二代彭布罗克伯爵（又名"强弓手"）缴械就范，而后者已在一个爱尔兰部落的指示下入侵爱尔兰并占据王位。这些英国士兵回国时报告了关于爱尔兰人消费"生命之水"的事情。这"生命之水"就是强劲的烈酒威士忌吗？如今爱尔兰威士忌的拥护者们对这种猜想表示赞同，不过还没有人能够证实。这些记录既没有提到他们喝的酒有何特征，也没有说明其生产原料。鉴于当时葡萄酒蒸馏技术在欧洲的广泛流传，这种酒很有可能是白兰地，而非威士忌。创作于 13 至 14 世纪的爱尔兰教会作品集《奥索里红皮书》中也出现了一份白兰地的酿制配方，这可以证实后一种猜想才是对的。

---

① 拉撒路是圣经中的人物，病逝后被上帝复活；该名字源于希伯来语，意为"上帝帮忙"。

然而几乎可以肯定的是，16世纪时爱尔兰人就在酿造类似威士忌的东西。当时谷物种植已普及，而且在1556年，英国议会在爱尔兰通过了一项法案，其中就担心地指出，因为"生命之水"，"人们每天都喝得醉醺醺的，而且爱尔兰王国境内随处可见人们饮用它的身影"。该法案允许社会精英们进行蒸馏活动，而其他人则必须向政府申请生产执照。

　　早期爱尔兰威士忌大都与早期苏格兰威士忌很相似，它们都是以大量发芽大麦为原料酿制而成的蒸馏酒，且风味浓郁。有些早期的爱尔兰威士忌蒸馏师确实会酿造"纯麦芽"威士忌，也就是完全使用发芽大麦酿造的威士忌。不过，如今大多数爱尔兰威士忌生产商已经采用发芽大麦和其他谷物的混合物做酿酒原料了。例如，有一份1873年的配方（或称麦芽汁配方）就要求以14%的发芽大麦、40%的未发芽大麦、16%的燕麦和30%的黑麦为原料。此外，直到19世纪中期，爱尔兰和苏格兰的蒸馏师都是使用铜质壶式蒸馏器来生产威士忌。一些早期的爱尔兰蒸馏师还会使用泥煤砖来烘干麦芽，赋予威士忌一种苏格兰式的烟熏风味。

　　然而到了20世纪初期，爱尔兰威士忌和苏格兰威士忌走上了不同的风味路线。最关键的是，爱尔兰蒸馏师不再使用泥煤。（如今，康尼马拉爱尔兰威士忌是唯一采用泥煤烘干工艺的爱尔兰威士忌品牌。）因此，尽管这两种烈酒都是以大麦为原料，但是爱尔兰威士忌发展成了一种散发水果香气且略带甜味的威士忌，苏格兰威士忌则继续保有或多或少的烟熏风味。

爱尔兰威士忌和英国的统治

1533 年，当亨利八世断绝国家和罗马教皇之间的关系，并自封为英国教会的宗教领袖之时，他也给自己制造了一个爱尔兰难题。四个世纪以来，爱尔兰名义上是处于英国的统治之下，但国王的权力却是来自 1155 年罗马教皇阿德里安四世颁布的任命状。为了不让爱尔兰的统治权回到罗马教皇或爱尔兰人手中，亨利和他的继任者们开始了一场漫长且残酷的斗争，以迫使这个国家臣服。英格兰派军队进入这个国家，并让英国新教徒和苏格兰长老会成员移居到爱尔兰的东北地区。

为了让爱尔兰人服从英国的统治，英国政府制定了多项政策，以减少威士忌的消费，控制爱尔兰人整体的饮酒情况。1580 年，当戒严令开始在芒斯特省实行的时候，英国人扬言要处死"叛军的帮凶"和"生命之水的生产者"。渐渐地，英国政府一步一步扩大了自己的权力。他们向供应麦芽酒和威士忌的旅馆征税，向威士忌生产商征税，甚至将用来生产威士忌的麦芽列为征税对象。个人如果想合法地酿造威士忌，则需要向当地政府官员购买专利证书或者垄断生产执照。也许最让爱尔兰人愤怒的是，从他们的酒中征得的税收，又变成拨给英格兰军队的经费。政府对威士忌酒征收的税越来越多，这对大多数贫穷的爱尔兰人来说负担尤为沉重。

不出所料，非法蒸馏在爱尔兰大行其道。爱尔兰人认为，与其购买更昂贵的合法威士忌（也被戏称为"议会威士忌"），倒不如喝私酿威士忌（单词 poteen 意为私酿威士忌，也拼作 poitin 或 potcheen），这样既可以省钱，又可以表达对英国统治的蔑视之情。合法烈酒的税费越高，非法蒸馏的"山间露水"①就越多。和苏格兰的情况一样，威士忌在爱尔兰也成了一项政治事业。一位姓名不详的诗人曾写过这样一首关于私酿酒的诗：

> 噢！创造私酿酒的人万岁，
>
> 教皇当然应该追认他为殉道者；
>
> 如果现在我是我们的女王维多利亚，
>
> 除了威士忌和水，我什么都不喝了！

最初，私酿酒只是一种未被征税的非法版爱尔兰威士忌。但是随着时间的推移，它逐渐演变成一种假冒的威士忌，甚至出现了更糟糕的情况。由于麦芽价格的上涨，私酿酒生产者转而选择糖、糖浆、土豆、大黄和苹果等更廉价的替代品。爱尔兰人想出了各种别出心裁的方法来藏匿他们的非法威士忌，所用容器的造型经过特别设计，便于藏在

---

① 原文为 Mountain Dew，美国俚语，指非法酿制或私卖的酒。

披风、大衣甚至女性内衣里。

英国当局最终意识到，实施越来越严苛的政策并非特别有效，于是他们便慢慢地通过修改法律、降低税费来鼓励合法生产威士忌。为了提高合法威士忌的声誉，政府在1759年通过了一项法案，禁止使用除发芽大麦、谷物、土豆和糖以外的任何原料来酿酒。

不幸的是，政府并不总能制定出明智的政策。1779年颁布的税法可能是最糟糕的一次决策。以往的威士忌蒸馏师是根据生产的烈酒数量来缴税的。显然，蒸馏师出于某些动机，会想办法让人觉得他生产的威士忌数量比实际的要少。政府对此的回应则是修改法律，改为根据蒸馏器的

1782年的一幅讽刺漫画描绘了政治家埃德蒙·伯克的爱尔兰性，画中同他一起出现的还有耶稣受难像、威士忌和土豆。

规格及其预期的月产量来征税。这个变化导致合法蒸馏师要么将自家的蒸馏器注销登记并且转入地下生产，要么加速蒸馏器的运转，以产出比政府预期产量更多的威士忌。在非法蒸馏快速发展的同时，议会威士忌的质量却大幅下降，这进一步促使了爱尔兰人选择喝私酿威士忌。

1783 年，政府扬言要对任何被发现存在非法蒸馏设备的城镇处以罚款，情况变得更加糟糕。这种集体惩罚政策不但没有推动社区对非法蒸馏施加压力，反而激起了整个社区的抵抗。在 1783 年政策出台后的 40 年里，经政府登记的蒸馏器数量从 1200 台降到了 20 台。

遗憾的是，为执行酒类法律而设立的政府机构不久就落得腐败且不幸的坏名声。19 世纪中期，税务机构尚未专

1816 年的一幅关于议会辩论的漫画，会上提到英国派去镇压私酿酒生产的士兵最终却变成私酿酒的饮用者。

业化，要成为一名税务官员并不需要对爱尔兰有所了解或具备蒸馏方面的知识。许多税务官员会向蒸馏师索贿，有些人这样做是因为本身贪污腐败，有些人则是因为政府发放的工资实在过于微薄。资产没收政策授权税务官没收任何用于生产运输非法威士忌的物品，包括蒸馏器、运货马车和马匹。市民和税务官员之间经常爆发冲突，有时英格兰还会派遣士兵去爱尔兰援助反私酿酒战斗。

爱尔兰威士忌的繁荣

英国政府合理地改进了威士忌政策，于 1823 年出台了适用于爱尔兰和苏格兰的《消费税法案》。威士忌生产商向政府登记蒸馏器的过程不再费力，而且政府只根据蒸馏师生产的威士忌数量来征税。威士忌法规和税收流程的专业化进程就此开始，并且在接下来的一个世纪里，税务官和蒸馏师之间逐渐形成了专业化的工作关系。更合理的法律和更专业的执法促使人们更加守法，也减少了人们生产私酿威士忌的动机。

此项政策变化的时机具有巧合性——威士忌产业的工业革命开始了。之前工匠们就一直在设计新机器，致力于将威士忌生产由一个粗糙的劳动密集型过程，转变成高效的机械化生产系统。尽管生产私酿酒可以获取一些利润，但是合法建立蒸馏厂并利用粮食收购和机械化生产的规模经济优势，却可以获得巨额财富。

爱尔兰威士忌超越了苏格兰威士忌，迅速发展为行业主流。例如 1825 年，米德尔顿蒸馏厂就建了一个容量高达 31,500 加仑的巨型壶式蒸馏器。与此同时，都柏林成了威士忌生产重地。约翰·詹姆森、威廉·詹姆森、约翰·鲍尔和乔治·罗[1]这些人的公司也都发展成了大型企业。乔治·罗的托马斯街酒厂占地 17 英亩[2]，每年的威士忌产量是 200 万美制加仑（750 万升）。在接下来的 75 年里，合法威士忌的产量翻了两番。

19 世纪爱尔兰境内的非法蒸馏器

---

① 四人统称为"四巨头"，是当时爱尔兰蒸馏产业的主导人物。
② 英美制面积单位。1 英亩约等于 4046.86 平方米。

1828 年，爱尔兰政客丹尼尔·奥康奈尔和威灵顿公爵"试图解决关于一滴爱尔兰威士忌的棘手问题"。

18 世纪中叶，一位到过爱尔兰的游客叙述了这样的情况：在一个小镇上，几乎每一家酒馆都会出售威士忌，当地人将其视为"有益健康且有镇静作用的利尿剂；他们总在吃饭前一起喝威士忌。为了让酒更美味，他们会把它倒进铁锅，再加入糖、薄荷和黄油"。加热一阵之后，把酒倒进罐子里，互相敬酒，一饮而尽。

在那个时期，除了爱尔兰，世界各地的顶级餐厅和俱乐部里也都有爱尔兰威士忌的身影。澳大利亚、印度和美国也都进口爱尔兰威士忌。爱尔兰北部科尔雷恩镇（位于德里郡）生产的爱尔兰威士忌还成了英国议会下院用酒。

物极必反

爱尔兰威士忌产业的发展并非没有代价,而其中最明显的代价就是酗酒成风,以及随之而来的健康和社会问题。在 20 世纪以前,爱尔兰还没有关于酗酒率的可靠数据。然而,当时有许多文字作品都记录了过量威士忌带来的不良影响。在这些叙述中,威士忌滥用这个主题总是和无端暴力交织在一起,令人担忧不已。

伟大的智者乔纳森·斯威夫特(1667—1745)曾描述了一场爱尔兰盛宴,宴上的"生命之水"是成桶成桶地供应。在那一晚结束的时候,情形莫名变得丑陋不堪。

> 是什么在刺什么在砍,
>
> 棍棒打得是多么响啊!
>
> 是什么直击内脏,
>
> 拳打脚踢得是多么狠啊!
>
> 一根根橡木棍,
>
> 被火焰烧硬,
>
> 一百个人脑袋破了,
>
> 一百个人腿瘸了。

与斯威夫特同时代的爱尔兰作曲家特洛·卡罗兰(1670—1738)的一首歌曲,展示了一个人与威士忌之间不健康的关系:

> 噢，威士忌！你可能会被大口吞下去，
>
> 是你让我的咽喉灼烧；
>
> 是你让我常常感到空虚，
>
> 金银离去。

这个嗜酒成瘾的倒霉酒鬼未能参加弥撒，为了喝酒连正餐都不吃，最终还失去了他的好名声。然而，他接着说道：

> 我确定我的衣服是你弄脏，
>
> 我确定我的鼻子是你割伤。
>
> 你把我打倒在地，
>
> 你让我失去活力；
>
> 但你总会在早晨治愈我，
>
> 从今往后我也原谅你。

这首歌的歌词可能是自传式的。作家奥利弗·哥德史密斯（1730—1774）曾说，卡罗兰"会整品脱整品脱地喝下生命之水"，而这个习惯"给他带来了一种难以治愈的疾病"，导致了他的死亡。

威廉·卡尔顿的《爱尔兰农民阶级的特征和故事》（1843年）描述了一种畸形的快乐，感受到这种快乐的是参加了残暴的帮派斗殴的天主教徒"绿带会员"①和新教徒"橙带

---

① 由爱尔兰乡村天主教徒组成的农民秘密会社，目的是以暗杀与纵火的方式剥夺地主的财产。

党员"[1]。对于他们来说，威士忌是点燃民族主义"热血"的"燃烧之酒"。下巴被打碎，脖子和骨头被打断，这一切都产生了奇怪的影响，让饮酒后过度兴奋的打手感觉"自己爱着所有的人，不管是男人、女人还是小孩"。

当然，正如经常发生的那样，爱尔兰人又一次做到了苦中作乐。民歌《芬尼根的守尸礼》可能就是最著名的例证：蒂姆·芬尼根每天早上都喝威士忌。一天早上，他出现震颤性谵妄，从梯子上跌落摔断了头骨。在他的守灵夜上，大家都喝得醉醺醺的，还爆发了一场争斗，而当"生命之水"溅到尸体上的时候，蒂姆死而复生了。

同样，19世纪一首匿名创作的歌曲《一口美味的威士忌》告诫人们不要滥用威士忌，但随后又调侃了一下不切实际的社会改良家。

> 一口美味的威士忌会让你欣喜；
>
> 太多的生物又会让你生气；
>
> 如果你理性饮酒，它让你开明；
>
> 如果你酗酒，它让你闭上眼睛……
>
> 一些医生会告诉你，它会损害你的健康；
>
> 法官会告诉你，它会减少你的财产；
>
> 医生和律师都十分认同，
>
> 如果你的钱都花光了，他们就拿不到报酬。

---

[1] 主张北爱尔兰继续隶属英国的新教政治组织成员。

然而医生和外科医师，

检察官和律师，

该轮到他们所有人喝一杯了。

爱尔兰威士忌的萧条

爱尔兰威士忌在 1900 年达到巅峰。那一年，爱尔兰境内 30 家蒸馏厂生产了破纪录的 990 万美制加仑（3750 万升）威士忌。当时的爱尔兰威士忌以高质量而闻名，二十年后这个产业却崩溃了。这是怎么回事？简而言之，是因为遭遇了太多坏事。有些本可避免，有些则无法避免。

19 世纪 40 年代，醉酒现象引发了社会的强烈抵制——禁酒运动开始了。几乎在同一时期，爱尔兰陷入大饥荒，由此引发的死亡和移民现象导致爱尔兰损失了近四分之一的人口（其中也包括饮酒者）。这两个因素使爱尔兰威士忌在本土市场的长期发展受到了限制。

尽管如此，得益于英格兰和海外的市场需求，整个 19 世纪爱尔兰威士忌的产量还是一直在增长。1870 年到 1900 年间，爱尔兰威士忌产量翻了一番，其中有 25% 至 60% 被用于出口。

不幸的是，爱尔兰并非威士忌市场上唯一的产酒国。如上一章所述，自 19 世纪 30 年代以来，在柱式蒸馏器的助力下，苏格兰蒸馏产业一直处于扩张的状态。位于都柏林和米德尔顿的大部分爱尔兰大型蒸馏厂都对柱式蒸馏器不屑一顾，认为其生产的是劣质威士忌。不过在北爱尔兰和苏格兰，柱式蒸馏器还是很快就被威士忌生产商所采用。

爱尔兰威士忌的巨头们向英国政府提出抗议，并多次请求英国政府出台法规，将威士忌和调和威士忌用不同的标签区别开来。他们声称，真正的爱尔兰威士忌需以发芽大麦为原料，并且经过壶式蒸馏器蒸馏而成。经过多番争论，1890 年，英国的国内外烈酒特别委员会只承认威士忌是"一种由酒精和水组成的烈酒"，这让使用柱式蒸馏器的生产商得以继续自由地生产谷物威士忌，并贴上"威士忌"的标签出售。

大多数爱尔兰蒸馏厂将新技术拒之千里之外，而他们也为这个决定付出了昂贵的代价，从中受益者是他们的竞争对手。苏格兰和北爱尔兰生产商在柱式蒸馏器的帮助之下，均在英国不断发展的二次蒸馏市场上占据了一席之地。（在 19 世纪，大量爱尔兰和苏格兰生产的威士忌都被运往英格兰，并通过二次蒸馏做成杜松子酒。）柱式蒸馏器的使用，让苏格兰和北爱尔兰蒸馏厂都能够给这个市场供应廉价威士忌。此外，这些威士忌生产商还在消费市场越来越受欢迎。他们和威士忌经销商合作，将柱式蒸馏器产出的威士忌，与风味浓烈的爱尔兰壶式蒸馏威士忌或苏格兰壶式蒸馏威士忌混合，生产出价格低廉、易于啜饮的调和威士忌。英格兰国内外的消费者很快就喜欢上了这些酒。

随着宗教和政治的干涉，"什么是真正的威士忌"这场争论变了味。许多爱尔兰调和威士忌都产自北爱尔兰。贪婪的北爱尔兰生产商被指责拉低了爱尔兰威士忌的品质，毁掉了这个行业。一个人杯子里的威士忌会被看作一种政治立场：买一杯产自都柏林詹姆森家或鲍尔森家的酒，你就是在支持新生的爱尔兰政权；点一杯百世醇，你就是在支持英格兰。

艰难时期开始于 1901 年。烈酒产量过剩，加之英格兰乃至欧洲普遍经济低迷，使得所有威士忌生产商的酒都滞销了。产量下滑，价格大跌。从 1900 年到 1915 年，爱尔兰威士忌产量从 990 万美制加仑（3750 万升）降至 700 万美制加仑（2650 万升），跌幅达 29%。爱尔兰蒸馏厂的数量则由 30 家缩减至 21 家。

第一次世界大战对爱尔兰威士忌产业造成了进一步的破坏。整个酒精饮品市场急剧萎缩，财政大臣大卫·劳合·乔治还利用战争推行禁酒政策。当产业工人因醉酒无法工作的故事流传开来时，乔治和其他社会改革家利用这些传闻，声称这些酒鬼是在破坏战局。

尽管英国的税收中有近 15% 来自烈酒产业，还有大概 30% 来自酒精饮品，但议会还是在 1915 年成立了（酒类交易）监察委员会。该机构有权采取措施，在生产或运输战争物资的地区限制酒类消费。委员会将这项职能的范围解释得非常宽泛，很快就在英格兰、威尔士和苏格兰施加了对酒类销售及消费的限制，这一限制最终覆盖了英国六分之五的地区。委员会还以"生理学原理""卫生原则"以及"国家效能"这个政府目标为名义，限制公众饮酒。唯一躲

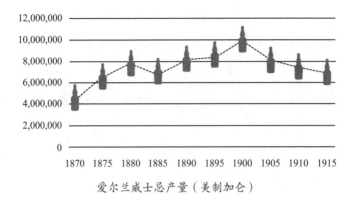

爱尔兰威士忌产量（美制加仑）

过委员会限令的只剩下英国偏远的乡村地区。除了这些严厉的措施之外，烈酒税被提高了近五倍，酒品生产也被严加限制，这就不难理解为什么会有那么多的蒸馏厂倒闭了。

　　与同样遭遇困境的苏格兰威士忌生产商不同的是，爱尔兰蒸馏厂缺乏足够的资金实力和海外市场份额来应对这些危机。爱尔兰蒸馏师固执地坚持使用大麦和壶式蒸馏器，于是也遗憾地错过了英国政府对工业酒精的巨大需求。1917 年，英国政府出台了大麦（限制）政策，这几乎给了爱尔兰威士忌生产商致命的一击。这项战时措施禁止大麦用于除生产食品以外的任何用途。多亏采用了柱式蒸馏器，大多数苏格兰威士忌厂存活了下来。1915 年，苏格兰威士忌产量跃升至 1850 万美制加仑（7000 万升），几乎是爱尔兰威士忌巅峰时期的两倍。由于苏格兰人种植和进口玉米及其他谷物来给蒸馏提供原料，大麦政策对他们的影响并没那么大。

爱尔兰和英国之间无法达成政治合作，给爱尔兰蒸馏厂造成了致命的伤害。爱尔兰自1916年开始反抗英国的统治，最终建立了脱离北爱尔兰的爱尔兰共和国。不幸的是，这场属于爱尔兰的胜利也点燃了贸易战，爱尔兰威士忌被英国拒之门外。

爱尔兰威士忌的销售还遭受到了另一个打击——1920年美国政府颁布的禁酒令。美国禁止威士忌进口，爱尔兰威士忌的巨大市场就此消失了十多年。更多的蒸馏厂关门歇业或进行业务合并以避免破产，威士忌产量也进一步下滑。

20世纪10年代晚期，都柏林詹姆森蒸馏厂外驮着一袋袋大麦的马车。

1933 年，废除禁酒令之后的美国市场重新开放，但爱尔兰的蒸馏厂由于缺乏足够的威士忌库存和产能，没能抓住飙升的市场需求。更糟糕的是，大部分现有的爱尔兰威士忌也无法出口到美国，因为 1926 年爱尔兰的一项法律将爱尔兰威士忌的陈酿时间从 3 年增加到了 5 年。其实这个想法是值得称赞的——通过规定一个比其他各类威士忌久得多的陈酿时间，使爱尔兰威士忌成为顶级精品。然而这个规定的短期影响是非常可怕的。美国还有一项法律规定：在生产国国内禁售的酒类产品，不得进口至美国国内销售。这意味着，陈酿时间低于五年的爱尔兰威士忌既不能在爱尔兰销售，也不能在美国销售。

涅槃重生

数十年来，爱尔兰威士忌产业萎靡不振，20 世纪上半叶的年产量可能只有 1900 年的十分之一。急需财政收入的爱尔兰政府一次又一次地提高威士忌的税率。从 1900 年到 1969 年，威士忌税增加了 2600%。

尽管如此，在 20 世纪中叶情况还是开始好转。爱尔兰政府选择将爱尔兰威士忌作为一种产品来支持，这样的支持也确实起了作用。1950 年政府确立了"爱尔兰威士忌"和"爱尔兰壶式蒸馏威士忌"的法律定义，这两种酒必须在爱尔兰境内使用发芽大麦为原料生产。后者必须完全使用铜质壶式蒸馏器蒸馏，前者则可以含有产自柱式蒸馏器

20世纪40年代，韦斯特米斯郡吉尔伯根镇上吉尔伯根（约翰·洛克公司）蒸馏厂外的男子。

的酒。政府还开始资助针对海外消费者的爱尔兰威士忌推广宣传活动。1964年，政府立法对产自爱尔兰境外的蒸馏烈酒征收关税，增加了爱尔兰威士忌对国内消费者的吸引力。五年后，爱尔兰威士忌的规定桶陈时间从5年降至3年，这有利于蒸馏厂更快地出售威士忌，并且桶内威士忌的蒸发量也更少了。

爱尔兰威士忌生产商也开始聪明起来。他们采用了柱式蒸馏器，产品更加多样化，生产出大量低成本的爱尔兰伏特加和杜松子酒。当时塔拉莫尔蒸馏厂生产的爱尔兰之雾利口酒，就是由爱尔兰威士忌、谷物威士忌、蜂蜜和药草调配而成的。厂商推广这款酒时将它宣传为一种古代饮品，而其最早的饮用者为一千年前的凯尔特人。爱尔兰之雾在美国和欧洲极度风靡。虽然蒸馏厂没能存活下来，但是爱尔兰之雾这款酒以及"塔拉莫尔露水"这个蒸馏厂著名品牌，依旧葆有着生命力。20世纪40年代和50年代，将爱尔兰威士忌与咖啡、糖和奶油混合制成爱尔兰咖啡这一创意传遍全球，让爱尔兰威士忌的销量更上一层楼。

为了利用规模经济的优势并减少爱尔兰威士忌蒸馏厂之间的竞争，1966年，仅存的几家公司合并成立了爱尔兰联合蒸馏者公司。（两年后，该公司更名为爱尔兰蒸馏者集团有限公司，简称IDGL。）昔日的竞争对手——约翰·詹姆逊父子、科克蒸馏公司和约翰·鲍尔父子，如今成了兄弟品牌。到了1975年，所有这些品牌都由米德尔顿一家拥有当时最先进技术的新厂生产。对于将购入的爱尔兰威士忌混制成调和威士忌售卖并把产品标上自家品牌名称的公司，IDGL不再向其售酒。IDGL统筹协调旗下所有品牌的生产和营销，给每个品牌确定了特定的市场。他们将未来押注在调和爱尔兰威士忌，而非纯壶式蒸馏威士忌上。

与此同时，北爱尔兰的蒸馏产业几近崩溃。它跟不上苏格兰蒸馏厂的步伐，且国内市场需求疲软。1947年，百世醇和科尔雷恩两家蒸馏厂建立合作关系。在这之后不久，老康博就关门停业了。1964年，科尔雷恩被并入百世醇，而它的蒸馏业务于70年代被关停。

20世纪70年代既标志着爱尔兰威士忌的终结，又标志着爱尔兰威士忌颇具讽刺意义的新开端。最初的时候，大多数爱尔兰蒸馏厂是由爱尔兰人经营的。很多情况下，公司的领导权都是代代相传，或者转交给管理能力值得信任

20 世纪 60 年代美国杂志上的百世醇广告

的家族世交。

这种做法止于 20 世纪 70 年代。百世醇被英国巴斯查林顿公司收购，而这家公司名下拥有许多酒吧。百世醇销量大增，这时候它那长方形的酒瓶出现在了英国各地的酒吧架子上。到了 1972 年，曾经难以想象的事情发生了——爱尔兰威士忌和北爱尔兰威士忌之间的古老裂痕闭合了。加拿大酒业巨头施格兰收购了百世醇，然后将其转卖给 IDGL，获得了 IDGL15% 的股份。至此，所有爱尔兰威士忌都齐聚 IDGL。

20 世纪末的爱尔兰威士忌

随着这个世纪逐渐走向尾声，爱尔兰威士忌也发展得越来越好。IDGL 对都柏林的老品牌尊美醇进行了重新设计，将其转变为一款清淡、顺滑并且果味浓郁的爱尔兰调和威士忌。IDGL 巧妙地对新版尊美醇和百世醇进行包装，并在接下来的 15 年里保持它们销量的稳定增长。以"三叶草和手杖"①的方式来销售爱尔兰威士忌的时代已经一去不复返了。那些品牌被重新定位为供世界各地人士消费的高档精致烈酒。

1988 年，一家英国公司热切地想要收购 IDGL，IDGL

---

① 三叶草和手杖均是公认的爱尔兰的象征，也是爱尔兰圣帕特里克节（爱尔兰守护神）的装扮象征；三叶草代表宗教上的"三位一体"的概念，手杖除了传统的拐杖作用外，也被认为是绅士解决争端的武器。

为了避开其收购，将股份出售给保乐力加。保乐力加是一家法国大集团，旗下拥有一系列多样化的饮料品牌，其中就包括必富达金酒和哟吼（一种不含酒精的巧克力牛奶饮料）。保乐力加将这个集团改名为爱尔兰蒸馏者有限公司，其威士忌销量得到进一步的增长。2007 年，尊美醇的销量增长了 15%，售出超过 260 万箱（约 650 万美制加仑或 2500 万升），而其销量于过去的一年中在保加利亚、俄罗斯、乌克兰和美国都取得了大幅增长，增幅分别为 61%、41%、43% 和 21%。

与此同时，2005 年百世醇蒸馏厂及其品牌以 3.24 亿美元的价格卖给了总部位于伦敦的帝亚吉欧；斯米诺伏特加、添加利杜松子酒、爱尔兰健力士啤酒和百丽甜酒都是这家公司旗下的品牌。帝亚吉欧正力图将百世醇发展壮大；据它公布的信息，2008 年百世醇在东欧的销售额就取得了两位数的增长。公司还计划，到 2011 年，要将年产量从当下的 95 万美制加仑（360 万升）增加到 240 万美制加仑（900 万升）。2010 年左右，百世醇推出了更多的酒款，如黑标布什、白标布什以及百世醇 10 年和百世醇 16 年。

爱尔兰威士忌的复兴让现在的消费者有了更多和更好的选择。20 年前，能在爱尔兰境外购买到的爱尔兰威士忌品牌只有一两个。而如今，有六七个甚至更多爱尔兰威士忌品牌可供选择的情况并不罕见，其中包括知更鸟这种纯壶式蒸馏威士忌，洛克这种单一麦芽威士忌，还包括顶级豪华威士忌，如每瓶售价超过 120 英镑或 100 美元的米德

2007年举办的吉尔伯根蒸馏厂创立250周年纪念仪式，现场还重启了已关停53年的蒸馏业务。从左至右依次是威士忌调酒大师诺尔·斯威尼、董事长约翰·蒂林和蒸馏厂经理布里安·奎恩。

尔顿珍稀。

还有一个迹象表明爱尔兰威士忌正走向复兴——爱尔兰威士忌的两大巨头得到了一个自视甚高的劳模的加盟：位于劳斯郡的库雷蒸馏厂。在短短的20年里，库雷就凭借其生产的威士忌轻而易举地赢得了一大堆奖章和奖项，基尔贝根、蒂尔康奈、康尼马拉泥煤等都是获奖酒款。2008年，库雷向40个国家输出了69万美制加仑（260万升）威士忌，并且还在不断增加自己的产量。

总的来说，爱尔兰蒸馏厂现在每年售出的威士忌有950万美制加仑（3600万升）——虽然与年销2.15亿美制加仑（8.14亿升）的苏格兰威士忌相距甚远，但仍是一个巨大的进步。

美国威士忌

在苏格兰和爱尔兰，英国的政府行为给威士忌生产商造成了巨大的影响，其政策对威士忌起到的作用既有正面的也有负面的。而美国的蒸馏厂就不会遭遇这样的困境。1782年英国结束对美国的统治，新的美国政府极少干预酿酒业。这种自由放任的环境促进了威士忌的繁荣，但是这种繁荣也在社会上引起了强烈的反对，导致政府颁布了禁酒令。最终，这个国家还是与威士忌握手言和，美国威士忌得以在全球市场迅速扩张，一飞冲天。

早期历史

美国联邦政府成立于1789年，但是在那之前，欧洲殖民者很早就开始蒸馏威士忌了。虽然在这方面权威的公共记录有限，但还存在大量其他零碎的证据。例如，1620年弗吉尼亚州一个叫乔治·索普的农民在他写给一个英国亲戚的信中提到："我们找到了一种用印第安纳玉米酿造好酒的方法，我好多次都拒绝喝美味强劲的英国啤酒，就是为了选这个酒喝。"1645年，弗吉尼亚州政府为白兰地和"生命之水"设置了价格上限，每加仑售价不得超过40英镑。

弗吉尼亚州往北400英里就是荷兰西印度公司的殖民地，也是现在纽约市的所在地。17世纪40年代的时候，那里的一家蒸馏厂可能就已经在用玉米和黑麦酿制私酒了。在弗吉尼亚州以南300英里的地方，也就是现在南卡罗来纳州的所在地，作家托马斯·阿什曾到此游历，并在他的

《卡罗来纳或彼国该州现状描述》（1682 年）一书中记录了殖民者喝玉米酒的情景。他们"发明了一种酿造优质啤酒的方法，但这种酒强劲易醉。将其浸渍再经过充分发酵之后，就可以在蒸馏器的帮助下从中提取出类似白兰地的强劲烈酒"。

情况跟爱尔兰和苏格兰一样，一些地方政治领导人担心农民会将大量用于生产面包的谷物转而投入威士忌生产，于是发布了法律和法院指令，暂时限制或暂停蒸馏行为。例如 1676 年，宾夕法尼亚州的一家法院只允许当地人使用"不适合研磨和煮食"的谷物来蒸馏。政府官员还对威士忌和其他烈酒给社会公德造成的影响感到不安。因为树立了坏榜样，牧师和宗教领袖受到指责。1631 年弗吉尼亚州的领导人宣布，"牧师不得过量饮酒，或酗酒狂欢"。

虽然美国在殖民地时期酿造了大量威士忌，但威士忌在当时并不是最受欢迎的酒。水果白兰地人们也喝得很多，但朗姆酒才是无可争议的王者。当时，美国把朗姆酒运往非洲，非洲把奴隶运到加勒比群岛的甘蔗种植园里干活，群岛再把用于蒸馏朗姆酒的蔗糖和糖蜜运往美国，这就是臭名昭著的"三角贸易"。

然而，朗姆酒位居美国顶级酒的时间并不长，取而代之的是威士忌。早期的美国是一个农业国家，广阔的土地上散落着许多规模较小的聚居地；玉米、黑麦和小麦在这个国家的大部分地区都长势良好。农民收获谷类作物，留足家用后把剩余的卖掉。谷物在市场上有时能卖个好价钱，

1736年，佐治亚州建立者詹姆斯·奥格尔索普看望居住在佐治亚州达里恩市的苏格兰高地人。此图绘于19世纪80年代。

有时则不行，但酒总是卖得不错。因此，农民和磨坊主都有强烈的动机生产威士忌，他们也确实这样做了。与此同时，随着这个年轻国家在陆路和航道上的发展，酒也可以被更轻易地运往远方的市场。美国威士忌产量飙升，售价下跌。1812年，新奥尔良港口见证了1.1万美制加仑（4.16万升）威士忌到港；四年后，到港数量达到32万美制加仑（120万升）。

与此同时，朗姆酒贸易却受到严重的干扰——国际奴隶贸易日渐式微（19世纪初期，美国和英国都开始脱离这项贸易），并且英美之间也展开了贸易战以及武装斗争（1775—1783，1812—1815）。此外，移民至美国定居的人们开始把朗姆酒和英国及令人深恶痛绝的英国皇家海军联系在一起。美国国内的支持者把威士忌吹捧为本土烈酒，消费者也乐于购买比朗姆酒（税费极高）更便宜的威士忌酒。到了1810年，在美国威士忌的消耗量已经超过了朗姆酒。

威士忌盛行、成为美国人首选酒的一个象征就是，美国第一任总统乔治·华盛顿，即"美国国父"，是一位威士忌爱好者。美国独立战争时期，乔治·华盛顿将军在指挥军队时曾恳求文职指挥官供应威士忌。"军队应该随时都能有充足的烈酒……以应付各种情况，譬如在高温或寒冷天气行军，驻扎营地或碰上雨天，身感疲乏或处于小组工作状态。此事至关重要，不可忽视。"华盛顿建议，政府应在各州建立威士忌蒸馏厂以供军需。他的提议无果而终。

结束总统任期（1789—1797）后，华盛顿到他的弗农山庄农场去过退休生活。他雇佣的苏格兰人詹姆斯·安德森力劝他建一座蒸馏厂，华盛顿照做了，一年之内他的蒸馏厂就生产了 1.1 万美制加仑（4.16 万升）的黑麦和玉米威士忌。在安德森的建议下，华盛顿还开始养猪，用糖化锅里剩余的谷物来把猪喂肥。

威士忌的繁荣

19 世纪是美国威士忌的繁荣时期。美国人口总数从 520 万增长到 7620 万；有了更多的人口和更多的农场，意味着会有更多的威士忌。技术创新也起到了推动作用——渴求利润的发明者为数十种新型蒸馏技术申请了专利，如通过蒸汽盘管加热蒸馏器。

此外，美国的税收政策也促进了威士忌的发展。简而言之，美国政府对进口酒征税，但很少对国产酒征税。这个年轻国家的政府吸取了 1791 年的教训：当年政府试图向国产烈酒征税，税务人员遭到酿酒商和酒客的攻击，次年还爆发了一场武装叛乱（即威士忌暴乱），乔治·华盛顿总统不得不派遣近 1.3 万联邦军队士兵前往镇压。直到 1862 年，美国才开始征收永久性的威士忌税。和英国一样，美国征税的理由也是战争——内战。

一般来说，美国政府视威士忌产业为合作伙伴，而非待宰的羔羊。威士忌的税通常都很低；联邦政府还允许酿酒商对放在仓库陈酿期间的威士忌免缴税款，这是一个非常明智的决策。这种"绑定"的做法始于 1868 年,并且蒸馏师可免税储存威士忌的年限逐渐延长,最高达 20 年。这一双赢的做法鼓励了威士忌生产商将产品陈酿,从而提升了威士忌的口感。对于陈酿期间蒸发的酒液,国会也给酿酒商免除了这部分税费(1880 年),并且通过了"保税威士忌"标签法案(1897 年)。此标签是质量的象征,因为它表明瓶中的威士忌是严格按照政府规定酿造的。与此同时,联邦政府的国库不断得到充实,威士忌税给政府提供了一半的收入。

与爱尔兰和苏格兰的情况一样,连续式蒸馏器在美国威士忌产业中也引起了混乱,由此引出了"什么是威士忌"这个古老的问题。没有法律规定禁止将少量优质威士忌与廉价的中性谷物烈酒混合,添加李子汁让酒液颜色变暗,再以威士忌的名义兜售。

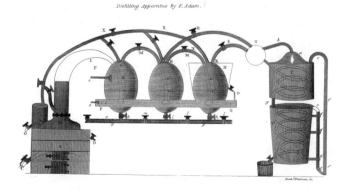

1800 年左右理查德·费尔曼绘制的新型威士忌蒸馏器

1794 年，一名被泼了焦油的收税官被威士忌暴乱参与者用横杆架走。此图绘于 1879 年。

1906 年颁布的《纯净食品和药品法》帮助解决了这个问题。该法规定，联邦政府有权要求商家保证消费品安全并贴上正确标识。西奥多·罗斯福政府（1901—1909）颁布法规，要求威士忌必需标有"调和威士忌""复合威士忌"或"仿制威士忌"。从那以后，这些酒标和定义都发生了变化，但宗旨不变：不要欺骗消费者——如实说出是什么样的酒。

物极必反

令人遗憾的是，早在威士忌出现之时，就有美国人滥用威士忌。1671 年的《纽约省执行委员会会议纪要》记录了一位外省官员提出的这样一条建议：

> 你们使用玉米来蒸馏烈性酒，既是谷物被大量消耗的原因，也是造成你们居民放荡和懒惰的原因，由此必然导致他们陷入贫穷甚至毁灭，此行为应该

被完全禁止或受到限制。

当时，部分殖民地已经颁布禁令，禁止使用烈酒和美洲印第安人做交易。但是这些禁令收效甚微。玉米酿造的廉价烈酒产量过剩，几乎到处都有在售，而且殖民者仍旧将其充当货币使用，美洲原住民饱受酗酒的摧残。休·莱弗莱在《当代人讲述的北卡罗来纳历史》（1934年）中记录道，1754年卡托巴部族的一位首领向美国人发出了请求：

> 兄弟们，有件事你们真该责备自己，那就是你们让那些谷物烂在桶里，然后又拿出来做成烈酒。你们还把这种东西卖给我们的年轻人，或者直接给他们……那些酒给我们的人带来了很糟糕的影响，让人内脏腐烂，病重危急。

正如历史学家威廉·J. 罗雷鲍所说，19世纪的美国变成了一个"酒精共和国"。1830年，15岁以上的男性和女性人均每年消耗9.5美制加仑（36升）烈酒。（如今，美国人均每年烈酒消耗量略高于0.7美制加仑，即2.65升。）当时廉价威士忌随处可见，美国人大量饮用。他们从早上就开始喝威士忌，有客人来访，有孩子出生，有人去世，无论何时都要喝威士忌。威士忌被人们当作药物，还被用于一些古怪的"疗法"。治疗浮肿时，痛苦的病人需要"日日夜夜"大口吞下混杂着芥菜籽、甜菜、辣根和褐色蛋壳的威士忌。如果病情没有好转，后续的治疗还是建议病人"重复同样的做法"。19世纪时，一位名叫安妮·罗亚尔的美国记者走访了美国中部地区的东边部分区域，对自己所看到的景象震惊不已："我在弗吉尼亚州的时候，那里的威士忌太多了……在俄亥俄州，威士忌太多了……在田纳西州，威士忌实在太多了！"

19世纪70年代的一幅偏执的讽刺画：一个微笑着的印第安人一手持来复枪，一手握酒瓶，威士忌从酒瓶里倾洒而出。

在选举期间，威士忌还扮演着一个不合时宜的角色。1758 年，乔治·华盛顿在弗吉尼亚州竞选议员时为选民提供了大量酒水，最后赢得选举；后来成为美国第四任总统的詹姆斯·麦迪逊曾因拒绝用威士忌收买选票，在 1777 年的州选举中落败。除了为选民提供一场免费的饮酒狂欢之外，赠送威士忌也被视为候选人关心普通民众的一种标志。

正如罗雷鲍阐述的那样，酒的过度消费导致了"家暴妻子、遗弃家庭和袭击事件，还有为酗酒者及其家庭提供援助的公款花费"。有些人对这种疯狂的行为嗤之以鼻。肯塔基州有个名叫小托马斯·约翰逊（1760—1820）的居民常被称为"丹维尔①的醉酒诗人"，他为自己的墓志铭创作了这样一首打油诗：

> 在这座大理石坟墓下，
>
> 无尽的阴影里躺着醉酒的汤姆；
>
> 安全停泊在这里，如木头般沉寂，
>
> 是谁饮下格罗格酒②后死去，
>
> 因威士忌格罗格而失去呼吸，
>
> 谁不愿死得如此甜蜜。

大部分美国人本质崇尚道德，对社会动荡不断加剧感

---

① 美国城市名。

② 即掺水烈酒，起源于英国。当年，英国皇家海军为掩盖存放太久的淡水的霉臭味，会向水中加入烈酒以改善口感，供水手饮用。为预防坏血病，还会在掺水烈酒中加入青柠果汁来补充维生素 C。

到不安。诸多大肆宣扬上帝、谴责烈酒的禁酒协会迅速成立。随着越来越多的美国人涌入这些组织，酒类消费量下降了。尽管如此，这些禁酒团体还在进一步加强他们反酗酒运动的力度。他们派发的数千万本宣扬戒酒的小册子里面还讲述了些有关酒精危害性的故事，但大都荒唐离谱。有个故事是这样的：一个酒鬼在一次工业事故中失去了一条腿，他把断肢卖给了一个声名狼藉的外科医生，然后拿着换来的钱继续狂饮作乐。还有些小册子利用了自由和自力更生这种美国式理想。饮酒被比作奴役，或是国王乔治统治下的"奴性"生活，人们劝诱酒徒放下酒杯，摆脱枷锁。酗酒者被指责为无法带来经济效益的公民。

1875 年，臭名昭著的威士忌酒帮丑闻进一步损害了酒徒们（后来被称为"湿派"①）的事业。大酒商贿赂政府官员，逃避联邦政府酒税高达数百万美元。尤里西斯·S.格兰特总统的私人秘书也受到指控，一场媒体与政坛的闹剧随之而来。

到最后，一些酿酒商和销售人员在推销威士忌时称之为包治百病的灵丹妙药，进一步玷污了威士忌的名声。1900 年 6 月 14 日，华盛顿特区的《晚间时报》用大块版面刊登了一则题为"治愈四百万病患——从未失败"的广告，宣称达菲纯麦芽威士忌已经治愈了 164,326 例痢疾、331,521例疟疾和 331,246 例"女性虚弱"。健康就在这一口酒之

---

① 禁酒令反对者，被称为"湿派"。

1867年美国南部非法生产威士忌的场景

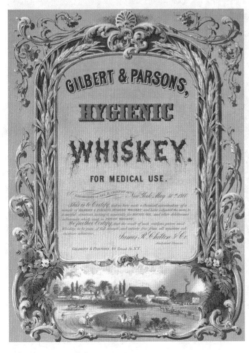

19世纪60年代的一则吉尔伯特和帕森斯牌"保健"威士忌广告

间——"如果你感觉身体有点不舒服，那是因为你血液失调。你需要一种刺激物。快按指示喝达菲纯麦芽威士忌。喝了它你就会痊愈，你的身体系统也不必遭受致命药物的伤害。"

一切已经注定——威士忌被认为与道德败坏、肮脏交易以及邪恶罪行紧密相连。

纳尔逊蒸馏厂为威士忌制作的古怪的土耳其式广告

禁酒令：美国的荒唐事

第一次世界大战期间，热爱自由的美国人为国家的利益做出牺牲，接受了政府的许多要求。言论自由受到限制，农业和工业产出由政府指导和配给。由于蒸馏厂都被用来生产供战争使用的工业酒精，酒类消费品产量大幅缩减。

随着"终战之战"的结束，禁酒团体做出了一个可能会让劳合·乔治拍手称赞的操纵策略——积极游说政府将战时的酒类管制措施变成彻底的禁酒行动。

对此，蒸馏师和酒客多年来一直在进行反击。他们多次组织反对抵制酒精的集会，并在报纸上谴责这些狂热的少数派侵害了他们的自由。乔治·加文·布朗（1846—1917）曾在肯塔基州路易维尔市创立威士忌公司，后来这个公司发展成为规模庞大的酿酒商——百福门公司。在其撰写的《圣经对禁酒令的否定》（1910年）一书中，他引用了圣经里对酒精饮料消费明确支持的内容，并宣称道："生产和销售醉人的烈酒，与生产和销售其他任何产品无异，并非道德更加败坏……人的每一个行为都要对上帝负责。"

这些都无济于事。那些被叫作"干派"[①]的人向联邦和地方的政客们递交了无数的信件和请愿书，最终说服他们通过了《禁酒法案》（1918年）和美国宪法第十八号修正案（1919年）。酒精饮料被列为非法事物，国家和州政府均有

---

① 禁酒令支持者，被称为"干派"。

权针对任何酒类生产者或交易者派出武装执法人员。

人们因禁酒令而难以获取合法酒，但还是能通过一些合法途径去要到一瓶威士忌。最能钻空子的方式就是以"医嘱"为借口。法律是这样规定的：

> 未经许可，任何人不得参与酒类的生产、销售、购买、运输或处方开具……有以下情况者除外，即虽自身未得许可，但持有医生开具的处方，则允许以药用目的购买和使用酒类……在一个十天疗程之内，为病人开具的口服处方不得超过一品脱纯烈酒，且配药次数不得超过一次。

美国作家哈里·克罗尔（1882—1967）发现了法律的漏洞，并在他的医生的帮助下拿到了酒。这个医生以治疗"感冒、流感、肌肉酸痛、便秘、腹泻、蛲虫病和痔疮"的名义，给他将酒开进处方。克罗尔一本正经地调侃道："有一次我用威士忌来治裆部瘙痒，结果发现口服药效更持久，也更美味。"为了满足克罗尔和其他酒鬼的需求，政府给美国药酒公司颁发了许可证。这是一个由肯塔基州六家蒸馏厂组成的财团，每年的波本威士忌产量令人瞠目，达140万美制加仑（530万升）。

禁酒令执行起来非常艰难。人们想出了各种巧妙的方式来隐藏威士忌，政府官员就曾发现，长方形瓶的尊尼获加红牌威士忌会被藏在长条面包里偷运。此外，美国大陆幅员辽阔，面积超过 300 万平方英里，所以在行政管理上，不可能完全禁止任何人在任何地方生产酒精饮料——要做到这点可能需要一支由数百万政府官员组成的军队，并且赋予他们几乎不受限制的搜查和扣押权力。

在 1921 年至 1923 年间担任美国总统的沃伦·G. 哈定曾对禁酒令嘲弄了一番。作为一个高尔夫狂热者，哈定会在高尔夫球场里喝威士忌鸡尾酒。首都还有许多人也违抗禁

禁酒令时期一瓶 1930 年左右的"药用"威士忌

酒令。黑麦威士忌被私酒贩从马里兰州带进首都；波本威士忌则在国会大厦一扇扇紧闭的门背后被大口饮下，而这栋楼还是美国国会所在地；走私贩搭乘小船或藏于载有其他物品的货船偷运爱尔兰和苏格兰威士忌，并将其带到盛行于纽约的地下酒吧；那些不希望受到指控的酒客则会登上从城市驶往公海的游轮，到了公海，威士忌和其他酒精饮料就都是合法的了。

舞女雷亚小姐展示她吊袜带上的一瓶威士忌。照片摄于 1926 年左右。

美国北部那些州的人还会越过边境去购买威士忌。当时的这首打油诗道出了他们对联邦当局不那么恭敬的态度：

> 二十四个北方佬口干得不行；
> 跑到加拿大去喝点黑麦威士忌。
> 酒一开，北方佬就开唱；
> 到底谁是柯立芝，天佑吾王。

当然，柯立芝是哈定的继任者，即试图让禁酒令奏效的美国总统卡尔文·柯立芝。加拿大威士忌不断流入美国。根据1930年2月美联社报道：

> 装着大量加拿大啤酒和威士忌的汽车从加拿大安大略省的阿默斯特堡出发，越过结冰的底特律河下游，开往国界线另一边的美国密歇根州……行驶过程中有一扇车门会一直打开着，所以当车驶过冰面的时候，司机便可以随时跳车。

加拿大于18世纪初开始酿造威士忌。加拿大的许多早期移民来自苏格兰，他们在伊利湖、安大略湖和多伦多附近区域扎根后不久，谷物磨坊和附设的蒸馏厂就出现了。和其他国家一样，加拿大在19世纪实现了蒸馏工业化，建立了规模庞大的威士忌工厂，如位于多伦多沿海地区的古德汉姆和沃茨蒸馏厂。美国的禁酒令试验至少有一个积极的影响——刺激了加拿大威士忌产业的发展。加拿大威士忌从两国之间绵长的边境线上倾涌而过。随着加拿大威士忌大获成功，海勒姆·沃克、约瑟夫·施格兰和塞缪尔·布朗夫曼等威士忌生产商发展成了行业巨头。加拿大俱乐部成为一个超级品牌，直到20世纪70年代，该品牌和其他加拿大威士忌都一直在《生活》和《花花公子》等美国主流杂志上刊登广告。

除了政策本身的愚蠢，禁酒令还造成了恶劣的影响。合法酒类销售额骤降，大批啤酒厂和蒸馏厂倒闭，美国政府损失了数十亿美元的税收收入。祸不单行，歹毒的罪犯通过生产和走私酒类发了财。毫无道德原则的犯罪团伙生产有毒烈酒，导致约4.5万美国人因饮酒而死亡，还有许多人遭受了永久性的神经损伤。

1922年华盛顿特区警方追捕威士忌走私贩之后的事故现场

幸而，1933年12月5日，美国政府结束了这场奇特的反酒精政策试验。全国各地的威士忌爱好者们鸣放礼炮、涌上街头，酒吧里人头攒动。税收收入源源不断地涌入美国国库，工人们也回到了饮料生产商的工作岗位上，这对于一个陷入大萧条的国家来说，无疑是两个积极的发展势头。

然而时至今日，美国人在酒精问题上的分歧依然存在。发生在位于田纳西州林奇堡市的杰克丹尼蒸馏厂的矛盾情况就是一个例证。

这家蒸馏厂于2008年卖出了超过220万美制加仑（830万升）杰克丹尼威士忌，多年来有数十万人慕名前往参观游览。然而，这群威士忌朝圣者中没有一个能在蒸馏厂里买到一瓶酒，因为这家蒸馏厂所在的穆尔县是一个禁止售酒的"干派县"。美国约有500个干派地区，穆尔县就是其中之一。与此同时，美国的50个州中仍有14个州禁止在"安息日"（星期日）销售酒类。

并且威士忌还是会被单独拎出来形容酗酒者普遍遭受的病痛——威士忌阴茎（阳痿）、威士忌大便（腹泻），还有威士忌脸（面部毛细血管破损）。没有人听说过伏特加震颤或者葡萄酒痴呆，而唯一和啤酒有关的贬义词就是"啤酒肚"。

长期以来，美国民谣、蓝调和乡村音乐固化了威士忌的糟糕形象。1950年，阿摩司·米尔本发布的流行歌曲《坏坏威士忌》就讲述了一个试图保持清醒的男人被威士忌征服、最终失去家庭的故事。同样，在《别再把威士忌卖给爸爸》（1954 年）中，一位父亲把家里的钱都花在纵饮威士忌上，导致整个家庭忍饥挨饿；歌曲的结尾是一个婴儿的痛苦哭号。还有些关于威士忌滥用的美国歌曲则是忏悔的自述：歌手坦白自己一个人喝得酩酊大醉，而这种情景通常是因为痛失所爱。约翰·李·胡克的《一杯波本，一杯苏威，一杯啤酒》（1966 年）和小汉克·威廉姆斯的《威士忌沉溺和地狱之旅》（1979 年）都是这种类型的作品。而在一些以张扬闻名的美国艺术家的包装下，威士忌的不良因素变得酷了起来，这也是他们"离经叛道"的生活方式所造成的一种影响。歌手贾尼斯·乔普林（1943—1970）的演出舞台上经常会摆两张凳子：一张自己坐着，另一张放着一瓶金馥力娇酒。迈克尔·安东尼（1954—）是范·海伦摇滚乐队中的一员，他弹奏用的低音吉他的外形与杰克丹尼酒瓶一样，

加拿大多伦多市的古德汉姆和沃茨蒸馏厂。照片摄于 1917 年左右。

有时他还会直接拿起杰克丹尼整瓶喝掉。作家亨特·S.汤普森（1937—2005）则时常把自己描述成这样：整瓶整瓶地喝下野火鸡波本威士忌，在肯塔基赛马会之类的活动上横冲直撞，开车时手里还抓着一大杯威士忌。

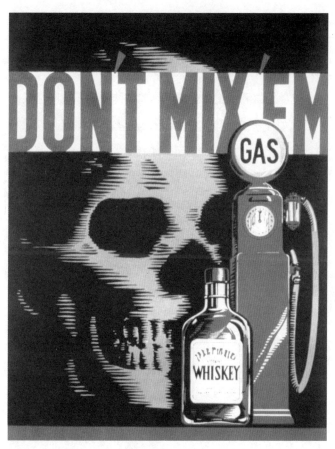

20世纪30年代中期，一张美国政府告诫民众不要酒驾的宣传海报。

艰难世纪，归于繁荣

禁酒令的废止并没有能够开启美国威士忌产业的好日子。肯塔基州有一半的蒸馏厂永久关闭了，全国能恢复经营的小型威士忌生产公司也很少。在美国加入第二次世界大战之前，这一萎缩衰退的产业只有十年的时间用于生产威士忌，战争开始后蒸馏厂便不得不转而生产用来制造迫击炮和其他战争物资的工业酒精。

此外，禁酒令还使美国酒客的口味转向了风味清淡的烈酒，如杜松子酒和伏特加，因为这些酒私酒贩可以很轻松地就大量生产出来。市面上能找到的威士忌通常都产自加拿大，这些酒由威士忌和中性烈酒混合而成，所以同样酒体十分轻盈。其他能找到的威士忌便产自苏格兰。这让美国那些急需资金的威士忌生产商不得不迎头赶上。为了重新获得市场份额，他们面临着一个选择：要么模仿苏格兰威士忌（几乎不可能），要么大量生产年份低且口味平淡的加拿大风格威士忌。他们选择了后者，而这几乎宣告了黑麦威士忌的终结——这可是一种从殖民时代就开始酿造的风格奔放的威士忌，带有胡椒味，香气馥郁，口感丰富。

回顾过去，美国蒸馏师决定追随潮流、放弃口味更粗犷的威士忌是可以理解的，但作为一项战略决策，这个决定还有待商榷。他们放弃了自己擅长酿造的酒，而热爱轻酒体烈酒的美国人还是继续购买加拿大威士忌，而且经常把酒兑成混合饮料大口饮下，比如 7-7 鸡尾酒（施格兰七皇

1950 年诺布溪蒸馏厂购自美国政府的蒸馏烈酒印花税票

冠威士忌和七喜）。除了加拿大威士忌，他们也会购买伏特加，而酒商推销伏特加时的卖点就是"干净"和"纯粹"，让人喝完以后不会浑身酒气。也有消费者热爱风味更加浓郁的酒，他们会购买苏格兰调和威士忌，而这类酒被营销为彰显酒客地位的高档进口商品。

20 世纪 80 年代，美国威士忌的情况开始好转。自朱莉娅·查尔德[①]开始尝试教美国人"掌握法国菜的烹饪艺术"之后，一场高质量饮食运动开始了。人们开始质疑为什么他们的食物多是预先做好的罐头食品，并且深陷于对更美味、更健康的食物的期待。长期以来因生产廉价罐装劣质酒而出名的加州葡萄酒商开始酿造世界级的葡萄酒；家庭式

----

① 美国知名厨师、作家与电视节目主持人，著有《掌握法国菜的烹饪艺术》。

作坊和精酿啤酒生产者也用了足量的麦芽和啤酒花来生产小批量啤酒，以此与米勒酿酒公司和安海斯-布希公司竞争市场份额。

美国蒸馏师的反应速度是慢了些，但当他们做出改变的时候，时机恰到好处。在20世纪60和70年代，酗酒和毒品泛滥问题对这个国家来说已是司空见惯的事。烈酒的人均年消耗量从1948年的0.7美制加仑（2.65升），增长到1978年的1.07美制加仑（4.05升）。（2006年的数据为0.71美制加仑，即2.69升。）那时的美国人拥有了比以往任何时候都多的可支配收入，并且也接受了"喝少点，喝好点"这样的理念。

由于产业整合，美国大部分威士忌生产转移到了肯塔基州和田纳西州。（就在禁酒令颁布之前，伊利诺伊州和印第安纳州的酒类产量比肯塔基州的还多。而60年后，肯塔基州的威士忌产量是印第安纳州的4倍，伊利诺伊州的15倍。）蒸馏师们将精力投入到两种不同风格威士忌的生产中——田纳西威士忌和波本威士忌。

前者只有两家公司生产，一家是规模小但质量高的乔治·迪科尔蒸馏厂，另一家则是旗下拥有杰克丹尼蒸馏厂的产业巨头百福门公司。正如第一章所提到的，田纳西威士忌与波本威士忌的不同之处在于，它在装桶陈酿之前会先经过木炭过滤。从1998年到2007年，杰克丹尼的销售额增长了65%。

相比之下，生产波本威士忌的公司有很多，几乎都位于肯塔基州。让黑麦威士忌和田纳西威士忌爱好者感到沮丧的是，越来越多的人视波本威士忌为美国国酒。从某种程度上讲，这完全是合理的——将玉米蒸馏成烈酒，再置于起源于美国的炭化橡木桶中陈酿。显然，波本威士忌产业支持这个观点，波本威士忌的拥护者也总喜欢指出，1964年国会通过的一项决议宣称，波本威士忌是"美国独具特色的产品……（它已经）得到了全世界的认可和接受"。这是事实，但只是部分的事实——整个决议并没有真的将波本定为美国官方认可的烈酒；更确切地说，通过这个决议是贸易保护所需。该决议总结道：

> 国会意见是承认波本威士忌为美国独具特色的产品，最后请美国政府相关部门注意，相关部门必须采取相应措施，禁止进口到美国的威士忌出现"波本威士忌"的命名。

尽管如此，美国国会参议院却于近期[1]宣布9月为"国家波本威士忌传统月"，波本威士忌也被其称为"美国本土烈酒"。

先不论官方的这一认可能带来什么好处，消费者显然对这一产品是非常满意的。自1999年以来，波本威士忌产量翻了一番，出口额从2006年的6.23亿美元跃升至2007

---

① 此书由原作者写于2010年。

年的 7.13 亿美元（上涨 14.4%）。这种增长很大程度上归功于"高端化"，即高价售卖更精致的威士忌。例如活福珍藏威士忌，从 2005 年至 2010 年其销售额以年均 24% 的速度增长。

美国所有的威士忌生产商都很明智地在向市场推出越来越多高质量的烈酒。除了老款黑标 7 号（22 美元），杰克丹尼目前还提供杰克丹尼绅士（30 美元）和杰克丹尼单桶（50 美元）供消费者选择。以白标波本威士忌（22 美元）闻名的金宾则推出了一系列在售的顶级波本威士忌，包括诺布溪（35 美元）、贝克斯（40 美元）、巴斯海顿（45 美元）和布克斯（60 美元）。规模较小的蒸馏厂也会售卖独具特色的产品，如老温克尔蒸馏厂就推出了酒龄 20 年的帕比·凡·温克尔家族珍藏波本威士忌（100 美元）。

而对于黑麦威士忌爱好者来说，现在似乎有了一线希望。多年以来，市场上只有老奥弗霍尔德和金宾这两家还在生产黑麦威士忌，而且还很难找到。而在过去的十年里，黑麦威士忌新品陆续上市，其中包括瑞顿房（15 美元）、6 年陈萨泽拉（25 美元）、老波特雷罗单一黑麦威士忌（60 美元）和 13 年陈温克尔家族珍藏黑麦威士忌（75 美元）。

21世纪的威士忌世界

21世纪的威士忌世界，是一个后现代的世界。曾经很简单的事——将谷物酿成酒，再主要卖给当地消费者——现在变得复杂到让人难以想象。威士忌生产已经全球化，数亿潜在消费者手握数十亿美元，准备购买产自世界各地的一系列威士忌以及威士忌衍生制品。对有些人来说，威士忌也已不再是一种简单的产品：它已然成了让人敬奉和崇拜的对象。

更多国家进军威士忌产业

苏格兰、爱尔兰、加拿大和美国并不是世界上仅有的威士忌生产国。半个多世纪前，日本突然进军威士忌产业，开始生产类似苏格兰威士忌的单一麦芽威士忌（日语名为Shinguru moruto）。山崎、响、白州、余市和竹鹤等一批日本品牌迅速涌现，而且质量普遍都很高。澳大利亚也是如此，现有维多利亚州的贝克里山蒸馏厂（成立于20世纪90年代后期）和塔斯马尼亚州的云雀蒸馏厂（成立于20世纪90年代早期），还有大南部蒸馏公司、赫利尔之路蒸馏厂等等。

这还不够，捷克、德国、新西兰、西班牙和土耳其近年来也都在生产威士忌。位于巴基斯坦拉瓦尔品第市的穆雷啤酒有限公司已经向市场推出以大麦为原料、经过壶式蒸馏的8年陈和12年陈的威士忌。1999年，在瑞典首都斯德哥尔摩以北140英里的地方突然矗立起一座麦克米拉

位于北海道的余市蒸馏厂。照片摄于 2006 年。

一瓶泰国的眼镜蛇
"威士忌"

威士忌蒸馏厂。该厂现在生产以大麦为原料的威士忌，工艺上会使用泥煤烘干大麦，并将酒液放在瑞典橡木桶中陈酿。

泰国一家公司曾一度销售一款以大米为原料的眼镜蛇威士忌，酒瓶里浸泡着人参、红辣椒和一条死去的小眼镜蛇，让人看了惶恐不安。烈酒评论网站（SpiritsReview.com）的克里斯·卡尔森称这款眼镜蛇威士忌有一种"辛辣的鱼腥味"，并且会导致口腔麻木和"四肢刺痛"。

全球化的品牌，纵横交错的威士忌关系网

假设一个多世纪以前，有一个叫麦克诺特的人开了一家名叫麦克诺特的威士忌公司，那它旗下应该也有个麦克诺特蒸馏厂，生产的酒便叫作麦克诺特威士忌并在当地市场销售。

随着全球化不断进展，公司、蒸馏厂、品牌和市场之间原本简单的关系也变得复杂了起来。美国标志性威士忌野火鸡波本威士忌就是这种后现代现象的一个佐证。

野火鸡威士忌的酒标上方印着"成立于1855年"，跟在下面的是"奥斯汀尼科尔斯"，再下面印的是"野火鸡"。一些消费者可能会这样解读这个酒标：自1855年起奥斯汀尼科尔斯蒸馏厂就开始生产野火鸡威士忌，并且大部分出售给了美国人。而事实上，位于肯塔基州劳伦斯堡（人口9014人）这座小城市的野火鸡蒸馏厂现在归意大利金巴利

集团所有，该集团于 2009 年初以 5.75 亿美元的价格从法国保乐力加集团手中收购了这家酒厂。野火鸡威士忌销往全球 40 个国家，销量达 200 万美制加仑（ 750 万升）。至于酒标中的"奥斯汀尼科尔斯"，那是一道历史遗留下来的痕迹——1855 年是奥斯汀尼科尔斯百货公司开门营业的年份①。

再举一个例子：尊美醇爱尔兰威士忌在哪家酒吧卖得最好呢？在爱尔兰科克市那座巨大的米德尔顿蒸馏厂附近？错，答案是明尼苏达州明尼阿波利斯市的一个名叫"爱尔兰本地人"的酒吧，但这个州只有不到 7% 的人口声称自己有爱尔兰血统。2008 年，这家酒吧的顾客共消耗了 1600 美制加仑（ 6000 升）的酒，换算下来大概每天 22 瓶，其中大部分是调成鸡尾酒喝掉的。

新个性化产品，新衍生品，新关联产品

全球化对于威士忌生产商来说是把双刃剑。服务的客户越多，潜在的利润也就越多。但是现在的市场竞争也是前所未有的激烈。没有哪家威士忌生产商可以高枕无忧，认定自家品牌可以长久地在酒吧和商店中占据一席之地。

---

① 奥斯汀尼科尔斯百货公司曾经有项业务是从市面上的蒸馏厂购进波本威士忌，再以自家品牌的名义包装售卖。据说公司的一名主管在 1940 年的一次野火鸡狩猎之旅中带了一些威士忌样品，深受大家喜爱，以至于后来朋友们向他要酒时都说"那个野火鸡波本威士忌"。奥斯汀尼科尔斯百货公司于 1942 年开始正式装瓶售卖野火鸡威士忌。

这种令人焦虑的环境推动了新威士忌以及威士忌衍生品的发展，而生产商瞄准的则是日益细化的市场需求。

独立装瓶的威士忌至少在 19 世纪就出现了，当时的商人先是购买酿好的威士忌，然后使用自己的品牌将其售出。如今，独立装瓶商不再对威士忌的来源遮遮掩掩，而是以此为卖点。他们的目标是吸引那些追求稀缺性的消费者，所以他们会从知名生产商那里购入威士忌，自己储存好，然后装瓶出售年份特别或强度特别的酒液。售价 100 美元的格伦泰特①就是个例子。麦卡伦向市场推出了很多款威士忌，唯独没有 19 年陈的，而这款特殊的苏格兰威士忌，只有在美国山姆会员仓储式商场才能买到。现在独立装瓶商的数量非常多，其中知名的就有高登麦克菲尔、默里麦克道维和斯科特。威姆斯年份麦芽威士忌是一家进军独立装瓶市场的新品牌，曾聘请著名威士忌作家查尔斯·麦克莱恩进行选桶。威姆斯威士忌单瓶售价通常都不低于 75 美元，在英国则至少 25 英镑。

正如我们所看到的，国家、民族认同和威士忌类型这三者之间的联系和威士忌本身一样历史悠久。现在我们谈论的依旧是爱尔兰威士忌和苏格兰威士忌，即使其生产公司并不在爱尔兰或苏格兰。

近来，附加在威士忌上的身份标识开始进入公众视野。JBB（大欧洲）上市股份有限公司（现更名为怀特麦凯有限

---

① 格伦泰特会向麦卡伦蒸馏厂购买威士忌装瓶出售。

格伦泰特 19 年麦卡伦单一麦芽苏格兰威士忌

责任公司）有段时间曾面向严守教规的犹太教徒生产弥赛亚威士忌。为了确保这款苏格兰威士忌符合犹太洁食认证的严格标准，当威士忌在苏格兰的因弗戈登镇进行混酿以及在格兰杰默斯镇装瓶时，都需要有一位拉比①在场。在大西洋彼岸也有类似的情况出现——老威廉斯堡 20 号是一款 3 年陈的波本威士忌，这个洁食酒款的名称来源于纽约市布鲁克林区的威廉斯堡，一个有许多犹太教正统派居住的街区。也许将来市场上还会出现越来越多的洁食威士忌。爱

---

① 犹太教经师或神职人员。

流行音乐家摇滚小子和纳斯卡赛车手罗比·戈登也是金宾波本威士忌市场营销的一部分。

尔兰的库雷蒸馏厂已经获得洁食认证，不过该公司尚未推出任何洁食酒款。

1996 年，昆廷·克里斯普单桶威士忌在英国发售，这出乎了很多人的意料。在此之前，几乎没有蒸馏烈酒生产商会面向同性恋消费者推出产品，克里斯普单桶是第一个迈出这一步的威士忌产品。虽然这个品牌转瞬即逝，但可以说它是威士忌产品在加强个性化和面向特定身份人群方面的先驱。

举个例子，美格波本威士忌就已经为肯塔基大学篮球队和肯塔基赛马会的粉丝以及一些慈善团体生产了限量版酒款。每位前往美格蒸馏厂参观的客人都可以通过亲笔签名和标注日期来定制专属个性化酒款，最后还可以手持酒瓶，将瓶颈浸入一桶热蜡中进行蜡封。为了满足特定节日或特殊日期（比如我们已经看到的千禧年主题威士忌）的关注者以及特定娱乐活动或事业的狂热爱好者，其他威士忌生产商推出特别版威士忌也属情理之中了。

许多威士忌蒸馏厂会向酒客出售期酒。当威士忌尚在橡木桶中陈酿时，消费者先付款认购酒桶，然后选定从橡木桶中取酒和装瓶的日期。有些蒸馏厂，比如华盛顿州的艾伦斯堡蒸馏厂，甚至可以将个人定制版的酒标贴到这类期酒的酒瓶上。

威士忌纯粹主义者的言行常常给人一种感觉，好像享用威士忌的唯一方式就是直接从蒸馏器或橡木桶里取出来喝。对于往威士忌里加其他任何东西的喝法，他们常常表示蔑视。但正如我们在第二章提到的，在一些关于威士忌的最早记录中，威士忌被描述为一种混合了草药、蜂蜜和其他东西的烈酒。这些年来，威士忌利口酒的人气有过鼎盛，也有过低迷，而现在它的春天可能又要到来。像杜林标苏格兰利口酒、爱尔兰之雾和百丽甜酒这些老品牌，在酒吧和商店的货架上一直占据着一席之地，而新的调配酒也陆续上市，比如凯尔特克劳斯爱尔兰威士忌利口酒和野火鸡美国蜂蜜利口酒。在 2009 年，比姆环球烈酒和葡萄酒股份有限公司曾推出一款红牡鹿波本威士忌，酒里加入了黑樱桃的风味。

同样，威士忌生产商在销售预调配制酒方面也取得了一定的成绩。这个想法其实很简单——提供单瓶调配好的鸡尾酒，帮消费者省下自行调配的精力。杰克丹尼可能是第一家尝试大规模生产和销售这种产品的公司，于 20 世纪 80 年代末推出了杰克丹尼乡村鸡尾酒（林奇堡柠檬水、黑杰克可乐等）。

威士忌作为一种产品，已经远远超出了瓶装酒的范畴。那些拥有狂热消费者基础的品牌已经生产出了关联产品，且产品形式早已不是那些在音乐会上分发或放在礼品袋里的廉价钥匙扣，也不是酒吧里的杯垫和促销酒具。威士忌爱好者对这些关联产品确实很买账，而且经常为它们花费不少钱。这些产品的出现并不是为了推广商家品牌，而是旨在利用消费者对品牌现有的强烈热爱，实现利益最大化。波本威士忌的超级生产商金宾就是这方面的高手。除了金宾莎莎酱和金宾牛肉干，粉丝还会购买各类金宾酱汁（烧烤酱、辣酱、腌酱、牛排酱和鸡翅酱）。这些还只是食品类商品，另外还有金宾服饰（T恤衫、拳击短裤和夹克）和家居用品（咖啡桌、导演椅、烧烤用具、壁挂式家用酒吧旋转灯等）。

敬奉和崇拜

为了满足威士忌信众，世界各地威士忌圣地不断涌现。肯塔基州莱克星顿市的蓝草酒馆向顾客提供 168 款不同品牌的波本威士忌；东京的橡木桶酒吧珍藏了大量极其稀有的苏格兰威士忌，其中包括年代久远的黑白狗和欧伯瓶装威士忌；苏格兰爱丁堡的凯德汉威士忌店有 200 多款苏格兰威士忌在售；芝加哥的宾尼酒水市场有500多款各种类型的威士忌可供选择。世界各地的威士忌蒸馏厂也纷纷敞开大门，建立游客中心迎接四方酒客。

在格雷厄姆·格林1940年的小说《权力与荣耀》中，读者们认识了"威士忌神父"，这个因该死的骄傲而背离上帝的神职人员，一步步走向放荡的深渊。当下的威士忌神父则更像是但丁笔下《神曲·地狱篇》中维吉尔这样的角色。一些诸如查尔斯·考德雷、约翰·汉塞尔、查尔斯·麦克莱恩、吉姆·莫瑞、加里·里根和加文·史密斯的作家会为初涉威士忌世界的新人介绍威士忌是什么、为什么和怎么样。他们的作品被刊登在《麦芽倡导家》《威士忌杂志》《波本国度读者》《威士忌生活》等杂志上，还会出现在众多网络出版物之中。酒客们如饥似渴地阅读他们的书，还常常在威士忌节举办之时蜂拥而至，就为聆听这些威士忌神父的长篇大论。

为了让世界上的威士忌爱好者可以将热情更充分地释放出来，以威士忌为主题的大型活动和体验活动越来越多。酒吧会举办品酒会，前文提到的威士忌神父中有些会组织私人样酒品鉴会，北美和欧洲会举办单一麦芽威士忌和苏格兰威士忌盛典、威士忌节和威士忌现场秀等聚会活动，每次活动都能吸引数百人参加。每年会有120多万人到苏格兰的蒸馏厂进行参观，还有来自十几个国家的约55,000名游客前往肯塔基州巴兹敦市（人口11,500人），只为参加当地一年一度的肯塔基波本威士忌节。

如今，虽然市场上还能买到 10 美元或 15 美元一升的威士忌，但它们已经被众多豪华威士忌挤到了一旁。一开始的冲击来自于单一麦芽苏格兰威士忌，接着是调和威士忌，现在则是躺在精美酒瓶里的爱尔兰威士忌、波本威士忌和黑麦威士忌，每瓶售价达 40 美元（或 30 英镑）甚至更高。

现在还会举办出售稀有威士忌的拍卖会，而且成交价常常超出人们的想象。爱尔兰戈尔韦市的修女岛蒸馏厂于 1913 年倒闭，而就在几年前，该蒸馏厂一瓶未开封的威士忌重现世人眼前，起拍价竟高达 15 万美元。众多蒸馏厂捕捉到这一市场趋势之后，从自家酒窖挖掘出越来越多的古老威士忌。大摩推出的水晶瓶装 50 年陈苏格兰威士忌，一瓶售价是 1.2 万美元，而产于 1926 年的麦卡伦珍稀系列单瓶售价为 3.8 万美元。为了吸引读者，威士忌出版物会经常刊登有关惊天价格的报道以及那些极其稀有的产品的照片，这就像是一种点燃读者激情的"威士忌色情片"，尽管大多数读者根本买不起或买不到这样的酒。

但也有极少数的幸运儿手里有着鼓鼓囊囊的钱包。世界的两端有两个人就因此而出名。意大利罗马涅大区卢果镇的瓦伦蒂诺·扎加蒂收藏了数以千计的苏格兰威士忌，它们的历史可以追溯到 19 世纪晚期。他的藏品这般令人惊叹，引得意大利图画出版公司为此发行了两本咖啡桌大小的影像集——《麦芽苏格兰威士忌最佳收藏》（1999 年）和《麦芽苏格兰威士忌最佳收藏 2：威士忌和更多威士忌》（2004 年）。

还有一个人在说英语的威士忌圈子里是无人不知的人

物，他就是哈维，据说是一笔巨额工业财产的继承者。在他名下位于华盛顿特区的巨型联排别墅里，这位威士忌爱好者在多个房间的墙面上摆满了威士忌。每隔几天就会有成箱的新货送到他家，哈维会煞费苦心地将它们分类，品尝后再进行评估。这些藏品的价值难以估量。有次我去拜访哈维，为了为难一下他，我便请他给我尝一下一种罕见的苏格兰威士忌。他带着我穿过一个又一个房间，几分钟内我们就喝到了那种威士忌的 14 个不同版本。他很慷慨地提出可以再多拿一些尝尝，但也提醒我道（他的做法是对的），如果一晚上将它们全都喝遍，可能会过犹不及。

在威士忌的硬核粉丝中，发展出了一股纯粹主义者形成的反对力量，这股势力虽小但日益增长，甚至带有些守旧主义。他们对今天所谓的威士忌不屑一顾，并且呼吁少些精致，多些"正宗"。我们从一些市场趋势就能看出这一点——"原桶强度"威士忌销量上升，并且从橡木桶直取、未加水稀释至 80 proof（酒精度 40%）或未过滤的威士忌也卖得更好了。

我们还可以从精酿威士忌的快速兴起中察觉出这股反力。和 20 世纪 70 年代末开始发展的精酿啤酒运动类似，精酿威士忌生产商也吹嘘自家的小产能和"旧世界"的工匠精神，并且以使用当地种植的有机谷物和使用壶式蒸馏器为卖点宣扬。美国一直是这场运动的温床。这类生产商有弗吉尼亚州的贝尔蒙特和瓦斯蒙德，加利福尼亚州的查贝和圣乔治，俄勒冈州的麦卡锡，科罗拉多州的斯特拉汉和纽约州的塔希尔。

在弗吉尼亚州的库尔佩珀县，查克·米勒坐在自己的蒸馏器上。图片摄于 2008 年。

令人惊讶的是，这场"回归本源"运动帮助了爱尔兰私酿威士忌重回市场。这一返场之路是由班拉提蜜糖酒和利口酒公司（位于克莱尔郡）与诺金山烈酒公司（位于沃特福德郡）引导的。这两家的私酿威士忌都未经陈酿，外观清澈如水。班拉提的酒有 80 proof（酒精度 40%），而诺金山的则有三种强度：120 proof（酒精度 60%）、140 proof（酒精度 70%）和让人目瞪口呆的 180 proof（酒精度 90%）。目前这些私酿酒的销量一般，但呈上升趋势。就在一个世纪以前，国王还曾动用军队镇压私酿威士忌的生产，然而在离奇的转折之下，现在人们已经可以在希思罗机场 3 号航站楼见到一瓶瓶正在销售中的诺金山私酿威士忌了。

威士忌生产商和公关公司都喜欢谈论"威士忌的美好旧时光"。如果这本书能说明什么，那应该就是现在正是威士忌酒客前所未有的好时光。经过政府的明智管控，经过资本主义的竞争，现在的威士忌比以往任何时代的都要优质。世界各地的消费者也可以买到更多更好的威士忌了：在南非的一家商店里，一位女士可以选购她想要的那个苏格兰威士忌品牌来招待她的客人；在巴西的一家酒吧里，一位坐着的男士可以通过网站 GiveReal.com，给他坐在芝加哥餐馆的朋友买一杯爱尔兰咖啡；在德国，一群好友可以商量着从一家苏格兰蒸馏厂购买一整桶专属于他们自己的酒。

让我们共举杯，为这琳琅满目的选择——敬威士忌，敬美好生活！

调酒配方

热托蒂

取大马克杯，倒入 1~3 盎司（30~90 毫升）任意类型威士忌。加入一片柠檬，蜂蜜的量根据个人喜好添加，再加入至少 5 盎司（150 毫升）的沸水。

爱尔兰咖啡

取马克杯或耐热玻璃杯，倒入 1~2 盎司（30~60 毫升）爱尔兰威士忌，根据个人喜好加入适量糖，再倒入 6 盎司（180 毫升）热咖啡。在顶上打上一层冻奶油或掼奶油。

曼哈顿 / 罗布·罗伊

取鸡尾酒调酒壶，装入三分之一壶的冰，倒入 2 盎司（60 毫升）波本威士忌或黑麦威士忌，以及 1 盎司（30 毫升）甜味美思酒。加入 3 抖振①的安高天娜苦精。摇晃均匀后倒入玻璃杯，再放入一颗马拉斯奇诺樱桃。如果是制作罗本·罗伊鸡尾酒，将基酒替换成苏格兰威士忌即可。

薄荷朱莉普

取坚固的玻璃杯，放入至少 3 片新鲜薄荷叶，倒入 3~5 盎司（90~150 毫升）单糖浆（即 50% 精炼白糖和 50% 水的混合液）。用碾槌将薄荷叶捣至出汁。往杯中倒入 2~3 盎司（60~90 毫升）美国威士忌或波本威士忌，搅拌，再用

---

① 鸡尾酒常用的特殊计量单位，酒瓶甩一次滴出来的量，大约是 1 毫升。

冰块填满玻璃杯（最好是刨冰或碎冰）。放一小枝薄荷作点缀。

苏格兰苏打

取玻璃杯，装入少量冰块，倒入1~2盎司（30~60毫升）苏格兰威士忌，以及6盎司（180毫升）苏打水。可以加螺旋状柠檬片作点缀。

威士忌可乐

取玻璃杯，装入少量冰块，倒入1~2盎司（30~60毫升）美国威士忌、波本威士忌或田纳西威士忌，以及6盎司（180毫升）可乐。可以加一片柠檬或一片青柠作点缀。

威士忌酸酒

取鸡尾酒调酒壶，装入三分之一壶的冰，倒入2盎司（60毫升）波本威士忌或黑麦威士忌。加入0.5盎司（15毫升）柠檬汁和0.5盎司（15毫升）单糖浆（即50%精炼白糖和50%水的混合液）。摇晃均匀后倒入玻璃杯，可选择是否加入少量冰块。放一颗马拉斯奇诺樱桃和一片橙子作点缀。

常见酒款推荐

美国 / 波本 / 田纳西威士忌

水牛足迹肯塔基纯波本威士忌

乔治迪科尔12号田纳西威士忌

老林头肯塔基纯波本威士忌

W. L. 威乐特别珍藏纯波本威士忌

加拿大威士忌

艾伯塔高级黑麦威士忌

加拿大俱乐部威士忌

皇冠威士忌特别珍藏调和加拿大威士忌

爱尔兰威士忌

1994年奈普格城堡单一麦芽爱尔兰威士忌

知更鸟爱尔兰威士忌

蒂尔康奈单一麦芽苏格兰威士忌

尊美醇爱尔兰威士忌

黑麦威士忌

金宾黑麦一号威士忌

老奥弗霍尔德黑麦威士忌

野火鸡黑麦威士忌

苏格兰——混合纯麦威士忌和调和威士忌

金铃喜乐顶级老苏格兰威士忌

芝华士

罗盘针享乐主义苏格兰威士忌

尊尼获加黑牌调和苏格兰威士忌

尊尼获加绿牌苏格兰威士忌

单一谷物威士忌

格里诺尔爱尔兰威士忌

苏格兰——单一麦芽威士忌

雅柏10年陈单一麦芽苏格兰威士忌

格兰威特12年陈单一麦芽苏格兰威士忌

拉弗格10年陈单一麦芽苏格兰威士忌

老费特肯10年陈高地单一麦芽威士忌

1994年格兰路思单一麦芽威士忌

致

谢

　　我要感谢瑞科出版社"食物小传"系列丛书的编辑安德鲁·F. 史密斯，是他给我写作此书的机会，他也是我的灵感来源之一。

　　衷心感谢瑞科出版社的迈克尔·利曼和玛莎·杰伊，他们用敏锐的眼光和敏捷的思维提供了编辑方面的协助；也感谢国会图书馆的 R. 山姆·加勒特博士，在我写作本书的过程中一直认真地聆听我的想法。

　　最后还要感谢以下各位为本书提供了数据、图表和资料方面的帮助：自由摄影师理查德·安东尼，泰勒公关公司的伊丽莎白·巴尔杜伊诺，葡萄酒交易所的詹妮弗·鲍林，烈酒评论网的克里斯·卡尔森，深蓝伏特加的罗宾·库帕，百福门公司的克里斯·休伊和沃尔特·特雷斯勒，顺风威士忌国际部的利斯克·拉森，殖民时代威廉斯堡基金会的玛丽安娜·马丁，保乐力加爱尔兰蒸馏厂的杰恩·墨菲和卡罗琳·贝格丽，智能通信公司的苏西·雷，酷雷蒸馏厂的杰克·蒂林和詹妮弗·格兰杰，科维斯通信公司的凯蒂·杨和切尔西·卡明斯，以及国会图书馆印刷与摄影部门无比乐于助人的工作人员。

　　书中任何不足和不妥之处，均是作者的责任。

Whiskey

⊖

A Global History

Kevin R. Kosar

Contents

Introduction

Whiskey is fascinating. There are hundreds of brands of it, yet few whiskies taste alike. Ardbeg and Laphroaig are both single malt Scotch whiskies. They are made in distilleries located a stone's throw from one another on the island of Islay, off Scotland's western coast. Yet their flavours are remarkably distinct.

Each whiskey reflects the ingredients, machines and people who make it. The passage of time wreaks changes small and large on all these factors; recipes get revised, water, soil and grain change, and both distilleries and whiskey-makers age and eventually are replaced. New brands of whiskey are born, old brands die, some brands even come back from the dead. So it is that a bottle of Bruichladdich Scotch whisky produced in 2010 will not taste exactly the same as one produced in 1950. For the curious soul, then, whiskey provides endless opportunities for discovery, as it ever evolves.

I also was drawn to writing this book because whiskey is not just a drink; it is a political, economic and cultural phenomenon. Arising sometime before the fifteenth century, whiskey began as an obscure and often throat-scorching drink that was produced by crude methods and enjoyed by poor village denizens and farmers in the British Isles. Today, whiskey comes in an astonishing number of types and brands, and it is consumed by the rich, the poor, the rural and the urban. Whiskey-making has moved far beyond its native shores and to places such as Germany, Japan, Pakistan and New Zealand, and whiskey-making has become as much a science as an art.

Does the world really need another book on whiskey? Yes. Here is why.

Most whiskey books are either tasting guides, or they address one type of whiskey (Irish, Scotch, bourbon, etc.) in its native land (Ireland, Scotland or the USA). These latter books tend to focus heavily on the stories of individual distillers and brands.

While this approach is illuminating and entertaining, it misses the bigger picture, and the remarkable parallels of nations' whiskey experiences: governments' struggles to regulate and tax it sensibly, and the sometimes violent responses these policies provoked; the booms, busts and industrialization of whiskey production; the politicization of whiskey; its seepage into nations' cultures; and the moral backlashes it has elicited. Additionally, examining all the world's whiskies together makes sense because today all whiskies are part of the global whiskey world and compete for consumers.

And I must also add that too many whiskey book

writers fall hopelessly in love with their subject matter. They uncritically repeat tall tales told by whiskey-makers and hooey put out by public relations firms, and they paint blissful scenes where crafty old-timers and strong-backed young'uns make whiskey 'the old fashioned way'. I have visited distilleries and whiskey towns; many are beautiful and populated with unforgettable characters. It is not for nothing that whiskey has been celebrated in poetry and song.

But the whiskey world also has mundane and ugly features. First and foremost, whiskey is a business. A few amateurs may make tiny batches for their own consumption, but most whiskey is churned out by computerized factories where professionals keep a sharp eye on the bottom line. Additionally, whiskey is a potent drink and many individuals cannot handle it. Whiskey misuse and abuse has sparked brawls, ruined families and killed many.

It is a wild tale that runs from the British Isles in the Dark Ages to the New World in the twenty-first century, and the plot includes political upheavals, technological revolutions, criminal enterprises, moral backlashes and globalization. The cast of characters is rich, with alchemists, crooks, eccentrics, poets, politicians, preachers, scientists and the millions upon millions who simply enjoy a stiff drink.

Chapter 1
Origins: From Seed to Spirit

Whisky vs. Whiskey

As with much in the whiskey world, there is a lot of nonsense propagated about the 'correct' usage of the terms 'whisky' and 'whiskey'. I have heard folks insist that 'whiskey' refers to the stuff produced in the United States. This is not true, as can be seen by looking at a bottle of the American-made Old Forester Kentucky Straight Bourbon Whisky.

Generally, in Canada, England and Scotland the preferred spelling is 'whisky', and in Ireland and the USA it is 'whiskey'. But plenty of exceptions to these rules exist.

In this book, the term 'whiskey' is used as a catch-all term covering both 'whisky' and 'whiskey'. The e-less form, 'whisky', is employed only when referring to the Scotch and Canadian versions, or to a specific brand, such as George Dickel #12 Tennessee Whisky.

'Whiskey' Defined

Put simply, 'whiskey' is an alcoholic beverage that is distilled from fermented cereal grain and aged in wooden casks. The term 'cereal grain' refers to those seed-producing grasses that we humans farm for consumption, such as barley, corn, oats, rye, wheat and so forth. Barrel-ageing is included as an essential part of the definition of whiskey because barrel-ageing imparts colour (from straw yellow to deep brown) and flavours (for example, vanilla) to the whiskey. Taken together, these two characteristics differentiate whiskey from other distilled spirits, such as cognac (which is distilled from fermented grape juice) and vodka (which can be distilled from grain, sugar beets or nearly anything else, but is not barrel-aged).

Like other liquors, most whiskies these days are bottled at 80 to 86 proof, which is 40 to 43 per cent alcohol. But not all of them. For example, George T. Stagg Kentucky Straight Bourbon Whiskey has been bottled at an eye-blurring 142.7 proof (71.35 per cent alcohol).

Whiskey-Making

Two processes are required to produce whiskey, *fermentation* and then *distillation*.

In simplest terms, *fermentation* is the production of alcohol from sugar, yeast and water. Beer and wine are produced through the fermentation of grains and grapes. In the case of wine, grapes are squashed to release their sugary juice, and yeasts either on the grapes or added by the winemaker feast upon the sugars and produce carbon dioxide gas and alcohol. With beer, an additional step is required. Before the yeasts can do their thing, the grains must be cooked in hot water to release their sugars.

Distillation is the process of heating a liquid to its boiling point, capturing its vapours and condensing them back into a liquid. So, then, whiskey-making might be conceived as the fermentation of grain and water into beer, and the distillation of that beer into liquor or distilled spirit.

Whiskey-making, then, is simple, right? Make beer, boil beer, catch its boozy vapours and let them condense into liquor—*voila!* Any idiot can do it!

Well, yes and no. Making foul-tasting, throat-burning bad whiskey is quite easy, but making delicious, sense-delighting good whiskey is extremely challenging. Many have tried to do it, few have succeeded.

This is because numerous factors can immensely affect the taste of the final product for better or worse. These factors include the character of the water employed, the quality, quantity and type of yeast(s) and grain(s) used, the shape, structure and mechanics of the still employed and the number of times the whiskey is distilled. After distillation, maturation carries myriad additional factors, such as the length of time the whiskey will be aged; the specific characteristics of the cask in which it is aged, including its size, the type of wood from which it was made, and the condition of the cask's interior (for example, new, fire-charred, or previously used to hold another beverage, such as sherry or bourbon); and the microclimate of part of the warehouse in which the cask is stored. Additionally, the whiskey-making process itself is inherently challenging; it is a marriage of chemistry and craftsmanship comprising four stages—grain preparation, fermentation, distillation and maturation and bottling—all of which must be carried out with great care in order to produce palatable whiskey.

Every distillery has its own methods and ways. What follows is a very simplified overview of the process for making whiskey.

Grain Preparation

In order to get spirit from seeds, the whiskey-maker must take a hard starchy seed and transform it into sugary material that can be fermented and distilled.

To understand how this works, we need first to consider a basic question: what is a seed? Put simply, a seed is a dormant, baby plant. On the inside is the embryo, which yearns to become a sprout; on the outside is the seed coat, a starchy layer that protects and nourishes the embryo. With the right environmental conditions (light, temperature) and water, a seed will awaken and begin to grow, producing enzymes (cytase and diastase) that help it convert the seed coat into carbohydrate (dextrin) and sugar (maltose) energy that the seed can use for growth. The whiskey-maker, meanwhile, wants the sugar for his own purposes. Accordingly, he will induce the seed to produce sugar but then stop the seed from using it.

To this end, the grain preparation process has three broad stages: *malting, milling* and *mashing.*

Malting

The malting stage produces 'malt', seed grain that can be milled and mashed, and then fermented and distilled into whiskey. Malting comprises three steps: steeping, germinating and drying.

The whiskey-maker steeps (or soaks) the grain seeds in water-filled tanks for a few days. Timing is critical; if the seeds are under-hydrated, they will not grow fully; if the seeds are over-hydrated, they will turn to mush. The whiskey-maker then removes the grain to large containers where they will start germinating (or growing). Their embryos awaken, and their metabolic processes commence.

Once the seeds have reached a precise point in growth, the whiskey-maker stops germination through drying (also called kilning). Hot air is blown over and through the seeds for a day or two. Those who have seen or visited whiskey distilleries may recall that many have small, pagoda-like tops. These tops vent the 60°C (140°F) air out of the distillery. How the drying air is heated is not an especially interesting subject, except in the case of Scotch whisky. Bricks of peat, decayed plant matter, are fed into a stove that heats the air. When burnt, peat emits smoke, seaweed and other robust aromas, which attach to the drying barley seeds. This is why Scotch whisky tends to smell and taste smoky.

Milling

The malted grain, or malt, now is ready for milling. To understand the importance of milling, it is worth recalling how bread is made. If one adds water and yeast to a bowl of rye or wheat seeds, nothing much will occur. If, though, one first grinds those rye or wheat seeds into flour, and then adds water and yeast, then one gets dough that may rise and be baked into bread. Roughly the same principle applies to whiskey. The whiskey-maker runs the sprouted seeds through a dressing machine, which snips the rootlets from the seeds, and then pours the seeds into a mill, which pulverizes them.

Mashing

Now, the whiskey-maker can extract the seeds' sugar, which he will later transform into alcohol. He does this through mashing—he dumps the milled mash into a large tank of hot water, called a mashtun, where mechanized rake-like arms churn the watery gruel. After 30 minutes or more, the water is drawn out through slits in the floor of the mashtun, and more (and possibly hotter) water is flooded in. This process may be repeated one to three times. Mashing activates the diastase enzyme, which converts the malts' starch and dextrin into sugar. The effect of mashing is remarkable—a starchy, unpalatable porridge becomes a sweet, drinkable liquid.

The most desirable portion of the mash, called the *wort*, is drained from the mashtun and run through a heat exchanger, a device that cools the sweet, steamy soup, as the wort flows into the washback, the large tank where fermentation occurs.

Fermentation

The star of the fermentation show is yeast, a single-celled microorganism from the fungus kingdom. The yeast's life is a simple one—it mindlessly eats, reproduces and dies. Well over 1,500 different species of yeast have been identified by scientists so far.

One species of yeast is especially favoured by alcohol-makers, *Saccharomyces cerevisiae*, which translates as 'sugar mould of beer'. This yeast thrives on sugars, which provide it with the fuel it needs to asexually reproduce—baby yeasts simply bud off mature yeasts, a bit like Athena popping from Zeus' head. With *S. cerevisiae*, in go sugars, and out comes alcohol, carbon dioxide gas and congeners, the latter term being a catch-all word used to refer to a host of acids and esters. Congeners are a mixed lot; some can contribute desirable flavours to the whiskey and others foul it.

Like other living things, yeast can only thrive within a limited temperature range, perhaps 10 to 37.8°C (50 to

100°F). Chilly temperatures stupefy yeast; high temperatures torment it. Pitching *S. cerevisiae* into boiling hot wort would kill it instantly, leaving one with nothing more than a great tank of cereal and water. (This is why wort is cooled in the heat exchanger.)

All yeasts work differently and produce different congeners. So distillers, like brewers and vintners, are extremely selective about the strain of *S. cerevisiae* they employ. Many whiskey-makers farm their own yeast colonies and store extra yeast colonies at a site away from their distilleries as a precaution. Using the exact same strain of yeast every time is a necessary step in assuring that the final whiskey tastes as the distillers intend and consumers expect.

The sugar-rich wort is a happy place for yeast. There it feasts and breeds. Fermentation has begun, and the washback of wort will bubble and foam for two days or longer, until the alcohol level of the wort reaches 5 to 10 per cent. Then the party is over; the yeast seize up and fall into a state of suspended animation. The result of this frenzy of activity is a mildly alcoholic brew called wash or distiller's beer.

Distillation

Despite its alcoholic content, the wash remains an organic substance. Within hours, air-borne microorganisms can invade it and instigate rot. To avoid this, the whiskey-maker hustles the wash into the still for boiling. Some distillers dump the whole of the wash into the still; others draw off only the most watery portion, leaving behind the yeast and the grain flour glop. It is a matter of the whiskey-maker's taste—whatever goes into the still will affect the flavour of the liquor that comes out of it.

While every still differs in shape and size, put crudely, there are two types of stills, pot stills and patent (or continuous or Coffey) stills. A pot still looks a bit like a giant copper gourd, with a bulbous base that narrows into a neck that bends sharply. When making whiskey, two or three pot stills are required. The patent still is a tall—9 metres (30 feet) high or more—set of two or more mechanical columns. Pot stills have been around for the better part of a millennium; the patent still did not come about until the nineteenth century.

In each case, the whiskey-maker uses the still to heat the wash sluiced into it. The temperature has to be just right—high enough to evaporate the alcohol in the wash, but not so high that it vaporizes the water in the wash. (The whiskey-maker is making whiskey, not distilled water.) Whiskey-makers also have to be careful that the heat applied does not scorch the wash, which would add unpleasant, burnt flavours to the spirit. The hot alcohol vapour rises upwards until it hits the 'condenser', the cool copper tubing (pot still) or plates (continuous still) that condense the vapour into liquid, distilled spirit. Copper plays a key role in the distillation process. Some of the undesirable compounds in the heated wash, such as sulphur, hit the copper and form a greasy compound, sometimes called 'grunge', that stays in the still rather than following the condensing spirit.

Not all the distilled spirit that stills produce tastes good enough to drink. The whiskey-maker must carefully monitor the distilled spirit as it flows. He does this by watching the spirit as it flows from the condenser. Instruments (often computerized) provide him with readings, but the whiskey-maker also may rely upon his own considered judgement of the colour and look of the spirit. Both the initial and concluding portions of the distilled spirit contain substances that are not tasty. So the whiskey-maker routes these foreshots and feints portions of the distillate back into the still, and directs the coveted middle cut of the spirit into the spirits receiver. A slight mistiming by the whiskey-maker can have negative effects on the whiskey which will not be detected until many years later.

Maturation and Bottling

The whiskey in the spirits receiver is clear as water and quite potent—140 to 160 proof. Usually, the whiskey-maker will reduce the proof of the whiskey by adding water to it before he pours the whiskey into barrels (or casks). Some distillers, however, offer their whiskies at 'cask strength', meaning that they do not water down the proof before bottling them. Thus, for example, Connemara Irish whiskey can be found in a cask strength of 120 proof.

Whiskey-makers and marketing people frequently talk up the importance of barrel-ageing and the particular qualities of the barrels they use and where they keep them. And rightly so. All the colour and perhaps 50 per cent of the flavour of a whiskey is produced through barrel-ageing. Once it is sealed in a bottle, whiskey changes little, if at all.

Barrels come in a range of sizes. Most American whiskies are aged in 53 US gallon barrels. Scotch whisky is stowed in 50 gallon barrels, 66 gallon hogsheads and 132 gallon butts. Size matters. The smaller the barrel, the more quickly the whiskey will absorb colours and flavours from it. This is a result of simple mathematics—in smaller barrels, the ratio of wood surface to whiskey is higher than in larger barrels. A hogshead, for example, has about 25 per cent more wood surface per litre of whiskey than a butt.

The type of wood and the conditioning of the cask also have profound effects on whiskey. Some whiskey rests in casks made from European oak; most, though, is aged in barrels made from American oak. Whiskey-makers are particularly fond of barrels made from oaks grown in chilly climates; their wood is denser since the trees grow more slowly. The denser the wood, the more durable the barrel, and the less whiskey will evaporate. Whiskey evaporation is no small matter; over a mere four years, a cask may lose up to 12 gallons of spirit. Multiply that by hundreds or thousands of barrels and you have fantastic sums of money disappearing into the atmosphere, which whiskey-makers call the angels' share.

Different types of whiskey are aged in differently conditioned barrels. US government regulations require bourbon whiskey to be aged in never-used white oak barrels, the insides of which have been charred by fire. Irish and Scottish whiskey-makers frequently age their spirits in used bourbon casks and Madeira and sherry wine casks.

The length of time that a whiskey is barrel-aged is driven by the judgement of the whiskey-maker, economics and the law. The whiskey-maker wants to optimize the flavour of his whiskey. A whiskey aged five years tends to taste better than one aged two years. However, each whiskey produced eventually faces declining and then negative returns from ageing. Thus it is possible that a whiskey aged 20 years may actually taste worse than if it had been aged only ten years. Additionally, the whiskey-maker also feels a financial imperative to get it to market for sale. It is true that older whiskey frequently fetches a higher price than younger whiskey, but the costs of running a distillery today must be paid today. Nations have helped define particular types of whiskey by establishing statutes and regulations that impose minimum ageing requirements. The US mandates that bourbon whiskey be aged not less than two years; Canada, Ireland and Scotland require their whiskies to be aged not less than three years.

Once aged, whiskey is drained from its barrels and routed into bottles. Vatted and blended whiskies are the exceptions to this rule. In both cases, the whiskey is splashed into troughs where it is married with other whiskies (vatted) or with other whiskies and grain whiskey (blended whiskey), grain whiskey being spirit produced from corn and other grains.

Lamentably, many whiskey-makers have taken to colouring their whiskies with caramel before bottling them. They have justified this practice by arguing that no two barrels age in precisely the same way, hence the colour of the whiskey coming from one barrel of ten-year-old MacCrusty Whiskey will not be the same colour as that coming from another barrel of ten-year-old MacCrusty Whiskey. To keep consumers from getting confused, caramel is employed to ensure that every bottle of MacCrusty looks exactly the same. Unfortunately, adding colour also feeds the popular misperception that the darker the whiskey is, the older and better it is.

Many whiskey-producers also run their whiskies through filters before bottling them. This removes bits of barrel char; filtration also removes fatty acids and esters, which can cause whiskey to appear cloudy when drinkers add water or ice to it. Critics have complained about this practice because they believe it sacrifices subtle flavours in the name of aesthetics.

The age listed on a whiskey bottle reflects the number of years that the whiskey has spent in a barrel. (How long the whiskey has been in the bottle is not reflected on the label.) A bottle of 'blended' or 'vatted' whiskey contains whiskies of different ages, and the age on the label must be the age of the youngest whiskey in the mix. As an imperfect rule of thumb, if the label does not state the whiskey's age, then it is likely that the whiskey was aged for less than four years.

The Types of Whiskey

A good way to give oneself a headache is to stand in the whiskey section of a liquor store or off-licence and begin reading all the bottles' labels. They carry a dizzying array of often impenetrable words that purport to differentiate the whiskey within from other whiskies: 'pure malt', 'sour mash', 'blended', 'small batch' and so forth. Some of these terms are useful; others obscure more than they reveal.

Plenty of whiskey dictionaries, such as Gavin D. Smith's *The A to Z of Whisky*, exist and define all these terms. Here we will stick to the basics. At bottom, what the person staring at the shelves of whiskey bottles wants to know is, 'What do they taste like?' One can get a rough approximation of the flavours a bottle holds by becoming familiar with the basic types of whiskey. One crude way for the novice to keep them straight is to categorize them by country: American whiskey tends to be sweet. Canadian whisky tends to be very light-bodied and fruity. Irish whiskey tends to be more robustly flavoured than Canadian whisky, and Scotch whisky tends to have a smoky flavour.

Once that is committed to memory, one can move on to the more elaborate definitions of whiskey and the various subtypes above, which are simplified versions of the complex definitions set by law and industry practice.

Importantly, these definitions provide clues as to the flavour of the product. Just as soup made from barley does not taste like soup made from corn, Irish whiskey (barley-based) does not have the same characteristics as bourbon (corn-based). Additionally, whiskies made with pot stills tend to be more robustly flavoured than whiskies made by patent stills (see chapter Four).

Clearly, making good whiskey is a complex process that requires a great deal of skill and practice. That anyone ever figured out how to do it, let alone do it well, is astonishing. Which raises the questions, 'Who invented whiskey, and when?'

Types of Whiskey

American Whiskey

Bourbon : Whiskey distilled from at least 51 per cent corn and aged in charred, new oak barrels.

Corn Whiskey : Whiskey distilled from at least 80 per cent corn.

Rye Whiskey : Whiskey distilled from at least 51 per cent rye and aged in charred, new oak barrels.

Tennessee Whiskey : Bourbon whiskey distilled in Tennessee that is filtered through charcoal and aged in charred, new oak barrels.

Canadian Whisky

Canadian Whisky : A blend of neutral spirit (which is nearly flavourless) and whiskey that is aged for at least three years.

Irish Whiskey

Irish Whiskey : Whiskey distilled from barley and possibly other cereal grains and matured in casks for at least three years. Most Irish whiskies are blends of Irish pot still, malt and grain whiskies.

Irish Single Malt Whiskey : Whiskey distilled at a single distillery from malted barley in a pot still.

Irish Pot Still Whiskey : Whiskey distilled from malted and unmalted barley in a pot still.

Irish Grain Whiskey : Whiskey distilled from barley and other cereal grains in a patent still.

Scotch Whisky

Scotch Whisky : Whisky distilled from barley and possibly other cereal grains and matured in casks for at least three years. Most of the Scotch sold today is blended Scotch.

Single Malt Scotch Whisky : Whisky distilled at a single distillery from malted barley in pot stills.

Single Grain Scotch Whisky : Whisky distilled at a single distillery from malted barley and possibly other cereal grains in a patent still.

Blended Scotch Whisky : A blend of one or more single malt Scotch whiskies with one or more single grain Scotch whiskies.

Chapter 2
Early History

Whiskey's Murky, Contested Origins

When was whiskey invented, by whom and why? These are straightforward questions, yet the answers are anything but certain. Depending on which book you consult, the answers vary, but usually involve either wise men in ancient Greece or the Near East discovering the process of distillation at least two millennia ago. Somehow, this knowledge made its way to the British Isles. Maybe St Patrick and Christian missionaries were the deliverymen. Or perhaps it was Islamic Moors who carried the knowledge to Europe. Proponents of the latter hypothesis like to point out that the word 'alcohol' came from the Arabic term 'al-kohl'. Once it arrived in the Isles, one reads that alchemists or holy men experimented with distillation and discovered that cereal grains can be made into a tasty spirit, one that ends up being called 'whiskey'.

As with anything else of value, there also are those who claim patrimony of whiskey for themselves or their home nation. Amusingly, some of these same persons will stake their claim while confessing they really have no clue. Witness Malachy Magee, author of the otherwise charming *1000 Years of Irish Whiskey*: 'Nobody can say for sure how or when it all began, but it is almost certain that whiskey, *uisce beatha*, or water of life, was first produced in Ireland.' How is Magee certain of this? He does not say. Magee alleges that Irish monks learned of distillation in their missions to the Middle East, and brought back the technique some time after AD 500 or 600. How exactly this adds up to '1000 years of Irish whiskey' is not clear, but what is indubitable to Magee is that the Scots are a wily bunch 'who proceeded eventually to capitalize on the bountiful gift of nature'.

Ralph Steadman, the impish Welsh artist, poked fun at this 'we invented it attitude' in his 1994 *Still Life with Bottle*:

> [T]he Egyptians in the times of Diocletian (AD 205–305) were really heavy boozers. They told the Babylonians who told the Hebrews. A heavily sedated foot messenger transmitted the news to Thrace, whereupon a traveling Celt peddling belt buckles and leather accessories...

Distillation in Western Europe, Steadman declares, was achieved first by—you guessed it—the Welsh, who in 1329 distilled lanolin, an oily extract of sheep wool, into a potent spirit.

The very notion that any one person may be credited as *the* inventor of anything is a dubious proposition. Working separately, Charles Cros and Louis Ducos du Hauron figured out colour photography; tinkering apart, Alexander Graham Bell and Elisha Gray concurrently devised the telephone. If those complex technologies can come into being concurrently, why not whiskey?

We Say 'Whiskey', But They Said . . .

The only ways that we today can know what people did aeons ago is to find evidence of their doings in writings and artefacts produced aeons ago. Without these, how would we know, say, that in AD 600 a hunched, hairy brute in Gaul discovered whiskey when he licked the dewy vapours of boiled, fermented barley juice that collected on the walls of his cave? The historian's job would be much easier if only the whole of human existence had been recorded on a computer hard drive. Everything anyone had ever done recorded and made full-text searchable—then one could find the answer to any question through the Googling of the proper words!

Yet, even if such a repository of human life existed, the historian researching whiskey would still be frustrated in his quest. This is because prior to 1900, there was little agreement on the meaning of the words used to refer to intoxicating beverages. There was verbal anarchy. In English language sources alone, one sees the word 'brandy', which today refers to a spirit distilled from fruit juices, used to refer to any liquor, be it made from fruit, root or grain. Similarly, the word 'liquor' was often used to refer to all intoxicating beverages, including beers and wines.

Thus far, researchers have dated the earliest appearance of the word 'whiskey' to a 1753 copy of *The Gentleman's Magazine*, which reported that in one Dublin shop 'there are 120 gallons of that accursed spirit, whiskey, sold'. Prior to then, the *Oxford English Dictionary* notes, there were mentions of 'whiskee' (also 1753), 'whisky' (1746), and 'whiskie' (1715).

The word 'whiskey' may have come from the word 'usky', (also spelled 'usquæ' and 'husque'), an apparent Anglicization of the Gaelic term *usque baugh*, which is pronounced 'oss-keh-baw', and which means 'water of life'. To add further uncertainty to the matter, *usque baugh* had at least a half-dozen other spellings, including 'usquebae' (1715), 'uskebath' (1713), 'usquebagh' (1682), 'uscough baugh' (1600), 'iskie bae' (1583) and 'uskebaeghe' (1581). The earliest known antecedent of the word is 'uisce betha' (1405), which apparently was pronounced differently by the Irish ('oss-keh baw') and the Scots ('ooshkie-bayha').

Yet in reading these early mentions in print of the word 'whiskey', one often struggles to understand just what the beverage is. The writers quite frequently did not describe or define the term, or used it to describe something quite different from the cereal grain-based drink that we call whiskey. A civil servant to the Crown in 1600 wrote of 'usquebaugh', a medicinal drink that was made with raisins. Likewise, a writer in 1658 mentioned a honey, wine and herb drink called 'usquebach'. George Smith in his 1725 *The Compleat Body of Distilling* gave a recipe for usquebaugh that included malted barley, molasses, cloves, coriander, cinnamon, nuts and sugar.

Thus, when we read in the *Annals of Clonmacnoise* (1405) that Richard Magranell, chieftain of the Moyntyreolas, croaked after consuming too much 'uisce betha', we cannot know for certain which sort of drink felled the man.

What (Little) the Historical Record Shows

So what does the historical record indicate about whiskey? Prior to 1500, the historical record is scant and awfully confusing.

The practice of fermentation goes back perhaps 10,000 years. During 8000 BC, grain was cultivated and processed in Egypt and the Near East. Archaeological evidence of these activities in Europe dates from 6000 BC.

There is solid evidence that the ancient Egyptians turned grain into beer and experimented with a variety of recipes for it. So too did the ancient Hebrews.

It is unclear whether Europeans got their knowledge of beer-making from the Middle East or if they simply figured it out on their own. There is evidence of beer-brewing in the UK in 3000 BC. Whiskey-making, though, would not come until many centuries later, when man married the techniques of both fermentation and distillation.

Interestingly, the Bible appears to differentiate beer and wine from something called 'strong drink'. Proverbs instructs: 'Wine is a mocker, strong drink is a brawler, and whoever it led astray by it is not wise' (20:1); and 'Give strong drink to the one who is perishing, and wine to those who bitter of heart' (31:6). Was this 'strong drink' a grain-based spirit? The Good Book just does not say.

A crude concept of the nature of distillation appears to have been known to the ancient Greeks. Aristotle's *Meteorology* (350 BC) includes this observation:

> Salt water when it turns into vapour becomes sweet, and the vapor does not form salt water when it condenses again. This I know by experiment. The same thing is true in every case of the kind: wine and all fluids that evaporate and condense back into a liquid state become water. They all are water modified by a certain admixture, the nature of which determines their flavour.

The most authoritative review of the existent historical evidence, R. J. Forbes's 1970 *A Short History of the Art of Distillation*, suggests that Egyptian chemists in Alexandria had worked it out in the first century AD or shortly thereafter. The writings of Maria the Jewess and others describe these early scientists' manipulations of liquids and substances— heating, pulverizing, mixing and filtering them. Renderings depict that they had the technology required to distil—a bulbous vessel (the cucurbit) for boiling a liquid that was topped by a narrowing and bent neck (the alembic) that condenses the vapours into a liquid that drizzled into a container (the receiving flask). The Egyptians, it is worth noting, had begun moulding glass vessels in 1500 BC, and by the first century AD had begun blowing glass containers.

Thereafter, the Arabs took distillation, which they viewed as a technique of transformation and purification, and turned it into a booming commercial industry. Arabic writings describe the recipes and techniques used to distil commercially desirable goods, including food additives (such as rosewater), essential oils and perfumes.

Over the next millennium, the knowledge of distillation and other early alchemical manipulations slowly moved both northeastwards and northwestwards. Along the way, the techniques were modified and put to additional purposes, such as producing crude herbal medicines. According to Forbes, traders and other travellers probably carried this knowledge from Alexandria to northern Africa and then Spain, and from Damascus to Constantinople (now Istanbul), and on to eastern and central Europe. By the twelfth century, Europeans were producing Latin translations of Arabic chemistry and scientific treatises. The English Arabist Robert of Chester, for example, spent time in Spain, where he translated *The Book of the Composition of Alchemy*, in 1144. He then returned to England, where he may have shared what he knew with others.

Nobody knows when someone first attempted to distil a liquid that already had been fermented. Aristotle wrote of evaporating and condensing wine. Edward Gibbon's *History of the Decline and Fall of the Roman Empire* (1776–89) mentions 'liquor' twice, speaking of fifth-century Tartars making liquor from horse milk, and Hun villagers making a whiskey-like spirit, 'camus', from barley. However, strong evidence of the production of distilled spirits does not appear until the fifteenth century. A treatise by Albertus Magnus (1193–1280), a philosopher and churchman, features a recipe for producing brandy that makes modern eyes pop:

> Take thick, strong, and old black wine, in one quart throw quicklime [calcium oxide], powdered sulphur, good quality tartar [potassium hydrogen tartrate], and white common salt, all well pulverized, then put them together in a well-luted cucurbit with alembic; you will distil from it *aqua ardens* [strong water] which should be kept in a glass vessel.

The Englishman Roger Bacon (1214?–1294?), and the Franciscan monks Raymond Lull (1234?–1315?) and John of Rupescissa (?–1366?) wrote of the power of distilled spirits to remedy physical corruption and to extend life, while the Florentine physician Thaddeus Alderotti (1223–1295?) and others devised improved methods for extracting pure alcohol from wine. The physician Arnaldus de Villa Nova (1235?–1311) had clearly discovered the power of distilled spirits—'It moveth some to excessive outbursts and some it leads to vigour and creative ecstasy—this water of life.'

Similarly, in Raphael Holinshed's 1577 *Chronicles of England, Scotland, and Ireland*, the Irish alchemist, poet and writer Richard Stanihurst (1547–1618) claimed that whiskey was a remarkable curative. 'Beying moderatelie taken, it sloweth age; it strengtheneth youthe; it cutteth fleume [phlegm]; it abandoneth melancholie; it relisheth the harte; it lighteneth the minde; it quickeneth the spirites.' If that were not enough, Stanihurst further claimed that whiskey cured dropsy, prevented kidney stones, released cramp-causing intestinal gas, steadied a queasy stomach and kept the circulatory system flowing and the bones sturdy. 'Trulie it is a soveraigne liquor if it be orderlie taken.'

The earliest irrefutable evidence of whiskey-making did not appear until the year 1494, when the Scottish Exchequer Rolls includes an entry reading: 'eight bolls of malt to Friar John Cor wherewith to make aquavitae.' It is worth noting that this is a hefty amount of malt, about 507 kg (1,118 lb), that would make perhaps 190 litres (50 US gallons) of booze, depending on the alcohol content.

Afterwards, mentions of grain being used to make distilled spirit appeared frequently in the Exchequer Rolls and the Accounts of the Lord High Treasurer. Obviously, it is entirely probable that someone was making whiskey before 1494, perhaps in Ireland, Spain or even the wilds of eastern Europe, as Gibbon claimed. Partisans of Scotland, though, may take pride that Friar John's 1494 entry is the earliest known reference, for now.

By 1506, whiskey was viewed with such reverence among the learned men of Scotland that the nation's King James IV, who enjoyed dabbling in alchemy and medical science, granted Edinburgh's Guild of Surgeon Barbers a monopoly in its manufacture. This government action is interesting on two counts. First, KingJames appears to have viewed whiskey as a curative. This perception of whiskey, and distilled spirits generally, as medicine rather than a recreational drink, was held by many of the well-read men of the day. This view of whiskey as medicine would hang on until the twentieth century (see chapter Five).

Second, the granting of this charter maintained, in effect, that government had the authority to decide who could and could not produce whiskey. In this case, the learned men of the Edinburgh barbers' guild were the benefactors of this largesse, and anyone of the city's poor who cooked up his own whiskey was in the wrong.

The Whiskey We Know vs. the Whiskey of the Past

Whiskey-makers and their public relations companies regularly boast of the ancient heritage of their whiskies. They claim they have been making their whiskey the same way for one, two or even three centuries. Their advertisements depict old men who carry on the heritage, using their burly, gnarled hands to distil whiskey according to a secret, immutable recipe. Not long back I attended a whiskey tasting where the representative of a well-known Scotch whisky company straightfacedly asserted that the whiskey his firm makes today tastes the same as it did 150 years ago.

Most of these claims are nonsense. What constitutes whiskey today is very different from what was called whiskey prior to the twentieth century. Over the past 125 years, governments have enacted laws and regulations that define whiskey's essential attributes, and set standards for the materials that may be used to make whiskey, what barrels it must be aged in and how long it must be aged. Before this happened, in the days of yore so celebrated in whiskey advertisements, rules were few and little enforced. Customers had no idea what was being thrown into the mash. It might be potatoes, sugar, oats, turnips or whatever else the whiskey-maker could get his hands on. Usually, it was not aged in barrels, and when it was, whiskey-makers used whatever barrels they could acquire, whether their interiors were new, charred or saturated with sherry, wine or the stink of pickled fish. Often, this so-called 'whiskey' was barely distinguishable from gin or liqueur, having been flavoured with honey, and herbs such as thyme, anise or mint.

The scant knowledge about the early history of whiskey is due, in great part, to the nature of whiskey production at the time. Whiskey was not a product marketed and sold for popular consumption. What evidence exists suggests that it was made by monks and alchemists, and was initially conceived as a medicine. It is entirely possible that whiskey also was made by farmers, those simple and often illiterate folk who lived outside the cities. Unsurprisingly, if they did make whiskey, they did not bother to record their achievements. Why would they?

As indicated above, it was not until government began to stick its nose into the business of whiskey-making that good documentation and records on whiskey-making appeared. As will be seen in coming chapters, James IV's grant of monopoly was an early shot in what would prove to be a long battle between government and individuals over the freedom to make whiskey.

Chapter 3
Whisky in Scotland

Generally, the story of whisky in Scotland is a happy one. The Scots appear to have made whisky first, but the Irish quickly began out-producing the Scots. By the middle of the nineteenth century, though, Scotland was on the rise. Despite temperance movements, economic depressions and world wars, Scotch whisky in all its diverse flavours and forms triumphed and is beloved the world over.

Early History

As we saw in chapter Two, the Scots were distilling spirits by the 1490s, and King James IV had granted Edinburgh a whisky-making monopoly soon thereafter. By 1579, distilling starchy plants into booze was widespread. The Scottish government worried that whisky-making might cause a famine as it consumed so much of the nation's grains. So authorities temporarily capped production by forbidding anyone but Earls, Lords, Barons and Gentlemen from whisky-making, and even they could only produce it for their own use. But whisky-making spread and, come 1644, the Scottish parliament saw wealth in whisky and began taxing it.

Whether whisky-making began amongst the learned in the city centres and spread outwards or vice versa is unknown. Regardless, stills were operating in the two centres of Scottish civilization, Edinburgh and Glasgow, and in remote farmlands and villages on islands to the north (Orkneys) and west (Islay). Martin Martin, author of *A Description of the Western Islands of Scotland* (1695), was taken aback by what he encountered:

> Their plenty of Corn was such, as disposed the Natives to brew several sorts of Liquors, as common Usquebaugh, another call'd Trestarig, id est, Acqu-vitae, three times distilled, which is strong and hot; a third sort is four-times distill'd, and this by the Natives is called Usquebaugh-baul...which at first taste affects all the members of the Body: two spoonfuls of this Last Liquor is sufficient dose; and if any Man exceed this, it would presently stop his breath, and endanger his life.

Whisky and English Rule

Many causes have been cited for Scotch whisky's generally happy tale—the Scots' thrift and pluck, topography (good soil, plenty of peat, many fine sources of water and so on) and Scotland's early embrace of capitalism.

An under-appreciated factor is the relationship between Scotland and England. Though it took two centuries, the Scots and the English were able to learn to cooperate politically and economically. This enabled the Scots to develop a whisky industry that grew and grew over the centuries. (Chapter Four on whiskey in Ireland presents a case in contrast.)

Serious cooperation began as early as 1603, when the two countries began sharing a king. Come 1707 the Union of the Parliaments folded the Scottish parliament into the representative body in Westminster, and formally intertwined the nations' economies. As part of this agreement, taxes on distilled spirits produced in the two countries were equalized.

Assuredly, this political cooperation of the Scots and English should not be overplayed—animosities remained. The Lowlands of Scotland were more congenial with the Crown than the Highlands (in the north). The former were Protestant and somewhat urbane, while many of the latter were Catholic and lived clan-based lives in rural areas. As an anony- mous bit of doggerel put it:

> O what lies younder north of Tweed?
> Monsters, and hillmen hairy kneed!
> And music that wad wauk the deid!
> To venture there were risky O!
> The fearsome haggis haunts the snaw
> The kelpy waits your banes to gnaw
> There's nocht to eat but oatmeal raw
> But still I'm told there's whisky O!

There were many clashes between the English and the Scots. Perhaps the most famous began in 1745, when 'Bonnie Prince Charlie', Charles Edward Stuart (1720–1788), led a Highland uprising in hopes of reclaiming the throne for his family. It ended harshly—many of his kilt-wearing troops were slaughtered at Culloden the next year, and the Westminster government subsequently forbade the owning of weapons and the wearing of kilts and 'Highland dress' by the Scots. (These bans were abolished in 1782.)

And whisky taxes were a sore point between the countries. Parliament first taxed distilled spirits in 1643, and whenever England found itself at war, it looked for money in whisky. It taxed the malt used to make Scotch, the stills, the spirit flowing from the stills and so on. In 1781 Parliament banned private distillation, and excise authorities were permitted to seize stills and any items used in the production or transportation of whisky, including horses and wagons.

The London government, to its credit, did learn the error of its ways. Parliament passed the Small Stills Act in 1816 to reduce whisky duties. Over the next decades, the law was amended to further lower duties on legally produced whisky, while the penalties on the production and consumption of illicit Scotch were jacked up.

But the government was not quite ready to trust distillers. It required each licensed distillery to provide space for a resident excise man, who would determine the amount of taxes that the distillery had to pay. The excise man measured the quantity of wash going into the still and the quantity of spirit ultimately produced, and generated a tax bill for the distillery. Parliament also mandated the use of a spirit safe. This sealed glass and brass box prevented the whisky-maker from dodging taxes by diverting whisky off the still before it could be measured by the excise man.

Dodging the Excise Man

In his *Inquiry into the Nature and Causes of the Wealth of Nations* (1776), Adam Smith saw clearly the problems with many of the government's policies: they 'made that a crime which nature never meant to be so'. Many people felt that the choice to turn grain into spirit was a private matter. It is not for nothing that folks began referring to whisky as 'innocent'. Moreover, many of the policies enacted took no account of incentives. As Smith pointed out, the more one taxes the distiller, the more incentive he has to distil illicitly.

At bottom, the authorities appeared to have been of the belief that they could control distilling by passing laws that told people not to do it or by taxing them heavily if they did. Obedience was initially assumed, and when Scots thumbed their noses at the law, armed government agents, the dreaded 'gaugers' or 'excise men' were loosed. Perversely, laws designed to increase tax revenues often decreased them as legal whisky production plunged and illegal whisky soared. In Edinburgh alone there were perhaps 400 stills in operation. Stills were cleverly hidden from excise officers— under bridges, beneath a home's floor (the steam and smoke were piped up the chimney), and even in a town's clock tower. Casks and jugs were buried in yards, tucked in trees and smuggled in caskets. According to lore, farmers in Oldbury, Gloucestershire used to hide their illegal whisky from government officials by keeping it in barrels marked 'sheep dip', a poisonous chemical used to keep bugs and fungi from attaching to sheep. Thus it is that even today one can find bottles impishly labelled 'Sheep Dip' in some of the world's better whiskey shops.

Rather foolishly, the authorities sought to reduce illicit production by offering cash rewards to anyone who handed over distilling equipment. Some whisky-makers turned this policy to their advantage. They turned in their worn-out stills and used the reward money to buy the material to make new ones.

No discussion of illicit whisky would be complete without mentioning those who brought it to market— smugglers. While many writers have made light of whisky smuggling, in truth, it often was an ugly way of life. Ian MacDonald, who served as an excise man, wrote of late nineteenth-century smuggling in the Highlands: 'I know most of the smugglers in my own district personally. With few exceptions they are the poorest among the people.' The smugglers' homes went unrepaired and their fields were little tended because they spent their days sleeping, tired from their nocturnal work. This lifestyle, MacDonald claimed, wore upon them, as did the heavy whisky drinking. In *Smuggling in the Highlands* (1914) he wrote that 'Gradually their manhood becomes undermined, their sense of honour becomes deadened, and they become violent law-breakers and shameless cheats'. Smugglers frequently assaulted and killed excise men for doing their jobs—hardly heroic behaviour.

Though often at odds, distillers and excise men sometimes came to mutually agreeable accommodations. A distiller might look the other way when the excise man took more whisky than was needed from the still. In return, the excise man might record the amount of spirit produced and subject to taxation to be lower than it actually was.

Of course, both distiller and excise man could make each other's lives very difficult. The excise man could write nasty things in his official reports about the distiller. In return, the distiller might torment the excise man. Since the latter was obliged to be on the scene when the distillation process began, the distiller might drive the excise man from bed by firing up the distillery at 3 am.

Over time, though, the relationship between the authorities and the whisky-makers improved as the government rethought its approaches to taxation and law enforcement. Newer policies employed incentives to encourage whisky-makers to carry on their trade legally rather than illegally. Incentives were altered to make licit production of whisky far more appealing than illicit production. Furthermore, the government also helped

Scotch-makers collectively improve the quality of their products by mandating good manufacturing practices, such as ageing whisky in casks. It was a win-win relationship, as distillers brought legal brands to market, government got its share of tax revenue, and drinkers had hassle-free access to fine-quality Scotch for the right price.

Thus, the armed clashes between excise agents and smugglers declined, as did the seizure of illegal stills. A productive working relationship bloomed. In 1983, excise officers ceased hanging about the distilleries. The duties for taking measurements and transmitting them to the excise office were handed over to the distillery manager. Today, the spirit safe is an artefact of yesteryear; it is no longer mandated by law. In short, as lawlessness declined and trust built up, the relationship went from one of policeman and suspect to that of external auditor and producer.

Scotch Whisky Booms and Booms

As the nineteenth century passed, the relationship between Scotland and England grew more cooperative. Some members of the English upper crust holidayed regularly in Scotland. Most famously, Queen Victoria and Prince Albert began summering at Balmoral Castle in the Highlands in 1848. There, they played at going native, sometimes wearing tartans and ordering casks of whisky from the nearby Lochnagar Distillery, which became 'Royal Lochnagar' after the Queen issued a Warrant of Appointment to John Begg, the distillery's head. (The Queen established no such residence in Ireland.)

As the British Empire expanded its global reach, Scotch whisky went with it, creating numerous converts. Production leapt as boatloads of whisky, especially blended malts, were shipped as far afield as the Bahamas, Egypt, India, Australia, New Zealand and South America. In the rapidly expanding and industrializing USA, drinkers also began to develop a taste for the slightly smoky whisky, despite America's own robust whiskey industry.

For all that, the first half of the twentieth century was brutal to the Scotch whisky industry. Lloyd George, the First World War, the Great Depression, the Second World War and subsequent rationing—these all knocked many whisky-makers out of business. The darkest days may have been in 1943, when no Scotch whisky was distilled at all.

Nonetheless, the Scotch distilleries lived on. Product diversification helped—with their continuous stills, Scotch distillers produced tens of million of gallons of industrial alcohol for the UK during both the world wars. (Ireland and England, meanwhile, had a retaliatory trade war in the 1930s, and Ireland remained neutral in the Second World War.) Additionally, the Scots benefited from their previous overseas agreements. Once the Axis powers were subdued and the world began to return to business as usual, the Scotch-makers began making and hustling their products to customers old and new. Japan represented a significant new market. Astonishingly, by the mid-1970s, over seven million US gallons (26.5 million litres) of Scotch were flowing into Japan each year.

Thus, while the Irish whiskey industry collapsed from 65 to three operating distilleries between 1875 and 1975, the number of Scottish operating plants actually grew from 112 to 122.

Ultimately, Scotch distilleries repeatedly expanded their plants to keep pace with the world's growing demand. Exports skyrocketed over the second half of the twentieth century, from 6.6 million US gallons (25 million litres) in 1950 to nearly 68 million (257 million litres) in 2000.

Scotch and 'Scottishness'

For all the nation's history, whisky has been just one of the many alcoholic beverages produced and consumed by the Scots. Beer probably came first, then brandy and liqueurs, and then whisky. Yet Scotch whisky became *the* drink of Scotland and part and parcel of Scottishness.

It is worth mentioning that were it not for bugs, Scotch might not have attained its high place in Scottish culture. Throughout what would become the UK, brandy (distilled spirit made from wine) was a very popular drink consumed by rich and poor alike until the late nineteenth century. Then Mother Nature intervened—phylloxera, a sap-sucking fly, began to blight France's vineyards in the 1860s. Wine production plunged, and brandy became very expensive and difficult to obtain. Scotch, meanwhile, was a bargain, and there was plenty to be had.

Initially, the fusion between the drink and the culture was produced by the rebellion against early parliamentary policies that targeted whisky. When the government taxed whisky and criminalized unlicensed production, many Scots dug in their heels. What had been just a drink became a cause.

Robert Burns (1759–1796) was Scotland's most famous early whisky publicist and romanticist.

> Let other Poets raise a fracas,
>
> Bout vines, and wines, and drunken Bacchus...
>
> I sing the juice Scots bear can mak us
>
> In glass or jug.
>
> O thou, my Muse! Guid auld Scotch Drink...
>
> Inspire me, till I lisp and wink,
>
> To sing they name!

Burns politicized whisky, melding it with Scottish identity and nationalism. 'Scotland, my auld respected mither!... Freedom and whisky gang thegither, Tak aff your dram!'

Writer Aeneas MacDonald (aka George Malcolm Thomson, 1899–2996) had a similarly romantic view of Scotch in his 1930 classic, *Whisky*:

> [There] dawned the heroic age of whisky, when it
> was hunted upon the mountains with a price on its head
> as if it were a Stuart prince, when loyal and courageous
> men sheltered it in their humble cabins, when its lore
> was kept alive in secret like the tenets of a proscribed and
> persecuted religion.

The notion that Scotch whisky was part of Scottishness was further fuelled by the whisky industry's marketing. The UK government stopped charging duty on advertising in 1853, and in 1860 it authorized the sale of spirits by the bottle. The Scots took advantage of these policies and began promoting their whiskies with gusto. They portrayed Scotch whisky as a gentleman's drink, something that persons of taste and distinction consumed. Scotch whisky promotions also carried ideal depictions of Scottish life that verged on caricature—kilted Scots hunting stags, playing golf, fly fishing and carrying bagpipes. Tommy Dewar began advertising his blended whisky with a bagpipe player dressed in a kilt and Highland-type dress in 1883, a practice that continues today. Brands with names such as Clan MacGregor, Clan Campbell and the tartan-labelled MacDugan reinforced this imagery. One of the earliest cinema commercials showed four kilted men dancing in front of a Dewar's banner.

Of course, not all ads were of this type. Some touted Scotch as a health tonic. Cambus Scotch called itself a 'wholesome stimulant' that 'ministers to good health and neither affects the head nor the liver'. Other promotional activities lurched into the outright ludicrous. Pattisons Ltd took parrots to bars where they screeched 'Drink Pattisons whisky!'

As the twentieth century progressed, the portrayal of Scotch as Scottishness shifted a bit. There was less earnest pastoralism, and more wit and mirth. Some advertisements played on the old image of the wily Scot who always manages to get his dram. The hilarious 1949 movie *Whisky Galore!* offers a similar picture. After a boat is wrecked off the Scottish coast, nearby residents try to snatch up its cargo of 50,000 cases of Scotch by outfoxing the authorities. No writer's fantasy, *Whisky Galore!* was based on a true story.

Meanwhile, in recent years The Keepers of the Quaich has tried to continue the association of Scotch and Scottishness while injecting a bit of dignity into the mix. This group extols the noble roots and respectability of Scotch whisky. (A quaich is a Scottish drinking cup that looks a bit like a bowl with handles.) Whisky industry folks founded the organization in 1988, though its style is decidedly old school. It adopted a coat of arms and a motto, 'Uisgebeatha Gu Brath', or 'Water of life forever'. It throws banquets at Blair Castle (in the Highlands, of course) and kilts are worn, bagpipes blasted, haggis forked and Scotch quaffed.

Too Much of a Good Thing

Of course, there were less happy depictions of the place

of Scotch in Scottish life. Thomas Crosland's *The Unspeakable Scot* (1902) is an infamous rant that disparaged Scotland for its cultural inferiority, moral corruption and overall 'mediocrity'. Crosland, who one contemporary described as a 'hysterically anti-Scottish' Englishman, howled at the purportedly high rates of criminality in Scotland, and claimed that the nation suffered from dipsomania. 'Scotland has become one of the drunkest nations in the world', Crosland huffed.

> Whiskey to breakfast, whiskey to dinner, whiskey to supper; whiskey when you meet a friend, whiskey over all business meetings whatsoever; whiskey before you go to the kirk, whiskey when you come out… whiskey when you are well, whiskey when you are sick, whiskey almost as soon as you are born, whiskey the last thing before you die—that is Scotland.

Crosland describes the typical Scot as careening violently from whisky-fuelled exhilaration to a 'dour' sobriety. 'You talk with him and get for answers grunts; he does not smile… He is glum, rude of tongue, and dull of mind.'

Later, less wild-eyed writers, such as George Douglas Brown, John MacDougall Hay and others, also wrote books depicting an ugly underbelly of Scotch whisky consumption—criminality, ruthless violence and personal destruction through alcoholism.

These depictions were not merely authorial fictions;

wherever one finds low-priced distilled spirits in abundance one will find unsettlingly high levels of alcoholism. Just as England suffered when dirt-cheap gin flooded its urban areas, parts of Scotland similarly succumbed to whisky.

In the 1830s Scots aged fifteen and older drank about a pint of licit whisky per week, in addition to illicit whisky and other alcoholic beverages. In parts of the country, whisky became embedded in everyday Scottish life. Whatever the occasion, wee drams of whisky were lifted. A wedding? Drink! A baby born? Drink! Somebody dies? Drink!

The general patterns of whisky drinking differed dramatically. In the rural areas of Scotland, there tended to be few public houses. In Shetland, for example, there was one pub per 1,000 people. Country folks, then, took whisky at home in small amounts throughout the day. A nip of whisky at dawn, another during a work break and a bit at dusk.

Densely settled urban areas were another story. In Glasgow, for example, there was one public house for every 130 people, and this does not count the illegal shebeens where cheap, white-hot whisky often called 'kill-me-deadly' was poured. Both the dreadful urban conditions and machismo encouraged binge drinking. The rules amongst tradesmen shops could be outright baroque. Drink funds were established, and workers could be tithed for both bad and good performance. Did the fire go out while you were tending it? Pay into the drink fund! Did you earn a pay rise? Pay into the drink fund! Once the pot grew large, the workmen would binge mightily, blowing it all on whisky in a night. Fishing and mining towns were especially whisky-sodden. George Bell, a mid-century anti-alcohol advocate, decried what he saw in the slums. 'From the toothless infant to the toothless old man, the population... drinks whisky. The drunken drama that is enacted Saturday night and Sunday morning beggars description. The scene is terrible.'

Happily, although the quantity of Scotch whisky produced skyrocketed over the nineteenth century, the Scots' consumption of it declined significantly over time. By 1900, the average adult Scot drank barely over a half-pint of whisky per week. Come 1940, consumption was down to a couple of sips of whisky per week.

Why the drop? It is hard to say. It has been suggested that the development of sports and recreation helped people fill their hours with something other than drink. Temperance societies began cropping up in the first third of the nineteenth century, and they probably diverted at least a few persons from boozing. Indubitably, a general lifestyle shift occurred, and spending one's nights getting hideously drunk and making a wreck of things became viewed as dysfunctional and lower class. (This shift also has occurred in Ireland and the USA.)

Additionally, governmental action had great effects. The Forbes-Mackenzie Act of 1853 greatly curtailed the hours that pubs could be open, and the Methylated Spirits Act of 1855 increased the government's powers to bust up shebeens and illicit drinking clubs. Slowly but surely, the government's measures reduced the times of day when whisky was available and the number of places that offered it. These policies, combined with gradual tax increases over the decades, increased the costs in money, time and effort that individuals faced if they were to acquire whisky.

Topography

Scotland is unique in the whiskey world in that its diverse topography has long made for diverse whiskies, especially those bottled as single malts. (Blended whiskies by their very nature lack these place-based differences.)

All Scotch whisky is made from water, peated barley and yeast. Yet a person who has never tasted Scotch could easily detect the difference between a Highland whisky like Glenlivet 12 Year, and an Islay whisky like Laphroaig 10 Year. The former is lightly flavoured, slightly smoky and shows floral and honey notes; the latter whisky floods the mouth and nose with smoke, iodine and seaweed flavours.

The diversity of Scotch whisky is the product of many variables, including different water sources, different types of barley, different yeast strains, different whisky recipes (mash bills), different distilling equipment and different climates in which the whisky ages. A few of these variables are local to the topography; the rest, though, are a matter of the whisky-maker's decisions. Nevertheless, the Scotch whisky industry still categorizes single malt whiskies in terms of five regions.

Islay whiskies tend to be the most flavourful malts and

typically smell and taste of brine, iodine and smoke. Some of the most well-known Islay whiskies are Ardbeg, Bowmore, Laphroaig and Lagavulin.

Contrarily, *Lowland* whiskies, those produced south of Dundee and Greenock, tend to be mildly flavoured and rarely have any of the Islay-type flavours. Only a few Lowland distilleries remain operating—Auchentoshan, Bladnoch and Glenkinchie.

Just to the east of Islay is *Campbeltown*, once a powerhouse of whisky production. Of the 30 distilleries that once operated there, only Glen Scotia and the much-revered Springbank distilleries remain active. Campbeltown produces Islay-like whiskies, though they tend to be less intense, and they show other light flavours.

Some of the most famous Scotch whiskies, such as The Glenlivet, Glenmorangie and Talisker, are made in the Highlands, an area situated immediately above the Lowlands. The distilleries there make an astonishingly diverse range of whiskies. Perhaps the only characteristics that Highland whiskies share are a tendency towards fruity flavours and a generally high quality.

Approximately half of Scotland's distilleries lie in the northeast of the Highlands, an area called the *Speyside*. Some of its enthusiasts proclaim it to be the crown jewel of the Scotch whisky industry. Indubitably, the Speyside produces many superb whiskies, including Aberlour, Balvenie and The Glenrothes. However, it is difficult to characterize Speyside whiskies as having a character that is distinct from

Highland whiskies. Indeed, Speyside whiskies themselves are remarkably diverse in character, ranging from the light and grassy Glenfiddich to the more robust and fruity Macallan.

Scotch at the End of the Twentieth Century

Scotch triumphed over the past century. Production soared, crossing 190 million gallons at the end of the century, and the selections multiplied. A dizzying array of terrific blended Scotch whiskies are available, from the low-priced White Horse and The Famous Grouse (c. £11 in the UK or US$18/litre), to the high-end Chivas Regal 18 Year Old (£50 or $50), and the super pricy Johnnie Walker Blue Label King George V (£535 or $400). Blends also are being produced at older and older ages. Cutty Sark, known for its bargain brand that was aged just a few years (£25 or $15), now offers twelve-, fifteen-, eighteen- and 25-year-old whiskies.

After decades of obscurity, vatted Scotch whiskies have begun to appear on store and bar shelves again. And single malts, which started the century as a distant second to blends, have become unbelievably popular. Distilleries have responded by offering more variations of their single malt whiskies, releasing them at different ages. In recent years, Talisker, for example, has offered malts at ten, twelve, eighteen, twenty and 25 years. Distillers also have experimented with ageing their Scotches in different casks to produce different flavours. Bowmore distillery, to name just one, has used casks that formerly held bourbon, sherry and Bordeaux wine. All this experimentation, of course, further blurs the differences between the characteristics ascribed to the particular regions' whiskies.

Chapter 4
Whiskey in Ireland

The story of Irish whiskey is a tumultuous one. From obscure, humble origins, Irish whiskey rose to commanding heights in the nineteenth century, only to suffer horrible setbacks that nearly spelled its end. Over the past half-century, though, Irish whiskey has experienced a Lazarus-like comeback. While there are far fewer distilleries and brands than there once were, the quality of today's Irish whiskey is superb and recognized worldwide.

Early History

Evidence of the distillation of liquor in Ireland dates to the twelfth century. In 1170, the soldiers of England's Henry II entered Ireland in order to bring to heel the second Earl of Pembroke (aka 'Strongbow'), who had invaded Ireland at the behest of one Irish tribe and then claimed Ireland's throne. On their return, the English soldiers brought reports of the Irish consuming 'aqua vita' and 'usquebagh'. Was this potent spirit whiskey? Partisans for Irish whiskey think so, though nobody can be certain. These accounts did not describe the characteristics of the beverage being consumed or indicate the ingredients used to make it. In light of the widespread distillation of wine in Europe, the odds would appear to be in favour of it being brandy not whiskey. The presence of a recipe for brandy in the *Red Book of Ossory*, a collection of Irish church writings produced between the thirteenth and fourteenth centuries, might be viewed as corroborating evidence for the latter hypothesis.

However, it is probably safe to say that the Irish were making something akin to whiskey by the sixteenth century. Grain cultivation was widespread, and in 1556 an act was passed by the English-run parliament in Ireland that noted with some concern that 'aqua vita' was 'daily drunken and used now universally through the realm of Ireland'. The act permitted the elite in society to distil, but it required everyone else to apply to the government for a licence.

In its early forms, Irish whiskey often was like early Scotch whisky. Both of these whiskies were robustly flavoured spirits made from goodly amounts of malted barley. Indeed, some early Irish whiskey distillers made 'pure malt' whiskies, that is, whiskies made entirely from malted barley. Most Irish whiskey-makers, though, have employed a mixture of malted barley and other grains. For example, an 1873 recipe, or mash bill, called for 14 per cent malted barley, 40 per cent unmalted barley, 16 per cent oats and 30 per cent rye. Additionally, until the mid-nineteenth century, both Irish and Scotch distillers used copper pot stills to produce their whiskies. And some early Irish distillers used peat bricks to malt their barley, which would have imparted a smoky, Scotch-like flavour.

By the start of the twentieth century, though, Irish whiskey and Scotch were on separate flavour paths. Critically, Irish distillers stopped using peat. (Today, Connemara Irish whiskey is the lone peated Irish whiskey brand.) So despite both being barley-based spirits, Irish whiskey developed into a fruity and slightly sweet whiskey, while Scotch continues as a more or less smoky spirit.

Irish Whiskey and English Rule

When Henry VIII severed his nation's ties with the Pope and set himself up as head of the English church in 1533, he created an Irish problem for himself. For four centuries, Ireland had nominally been under English rule, but the Crown's power came from a writ issued by Pope Adrian IV in 1155. Rather than let control of Ireland revert to the Pope or to the Irish, Henry and his successors began a long and often brutal effort to force the nation into submission. England sent military forces into the country and populated its northeast with Protestant English and Presbyterian Scots.

As part of the effort to get the Irish into line the English government enacted multiple measures to reduce whiskey consumption and Irish drinking generally. When it imposed martial law in Munster in 1580, the English threatened to execute the 'aiders of rebels' and the 'makers of aqua vita'. Bit by bit, the Crown extended its power. It taxed inns that served ale and whiskey, and it taxed whiskey producers. When individuals sought to make whiskey legally, they had to purchase a patent, or monopoly producer licence, from a local government official. Even the malt used to produce the whiskey was taxed. Perhaps especially galling to the Irish was that their drinks were being taxed to finance England's army. The imposition of greater and greater duties on whiskey were especially burdensome to the mostly poor Irish.

Not surprisingly, illicit distilling boomed in Ireland. Rather than purchase the more expensive legal whiskey, derisively called 'parliament whiskey', the Irish could save money and thumb their noses at British rule by drinking poteen (also spelled poitin and potcheen). The more spirit taxes were increased, the more 'Mountain Dew' flowed from illicit stills. As in Scotland, whiskey became a political cause. One unknown poet wrote of poteen,

> Oh! Long life to the man that invented poteen,
> Sure the Pope ought to make him a martyr;
> If I myself was this moment Victoria, our queen,
> I'd drink nothing but whiskey and water!

Initially, poteen was just an untaxed, illegal version of Irish whiskey. Over time, though, it morphed into a pseud-whiskey or worse. As the cost of malt increased, poteen-makers turned to cheaper alternatives, including sugar, treacle, potatoes, rhubarb and apples. The Irish devised ingenious means to hide their illicit whiskey. Containers were fashioned that could be easily hidden under cloaks and coats and even women's undergarments.

Eventually, English authorities recognized that imposing increasingly draconian measures was not particularly effective. Slowly, the laws were revised and taxes were lowered to encourage legal whiskey production. In hopes of improving the reputation of legally produced whiskey, the government passed an act in 1759 that forbade the use of any materials other than malted barley, grain, potatoes and sugar.

Unfortunately, the government did not always make smart policies. Perhaps the worst tax enacted came in 1779. Previously, a whiskey distiller was taxed based upon the quantity of spirit he made. Obviously, the distiller faced some incentive to make it appear as if he was producing less whiskey than he actually did. The government responded by changing the law to base the tax upon the size of the still and its expected monthly output. This change encouraged the legal distiller to either deregister his still and take production underground, or to run his still more quickly in order to produce more whiskey than the government thought he could produce. While illicit distilling skyrocketed, the quality of parliament whiskey plunged, further encouraging the Irish to drink poteen.

Matters were made even worse in 1783, when the government threatened to fine any town where illegal distillation equipment was found. Rather than encourage community pressures against illicit distillation, this policy of collective punishment galvanized community resistance. In the forty years succeeding the enactment of the 1783 law, the number of government-registered stills fell from 1,200 to twenty.

Regrettably, the bureaucracy created to enforce the liquor laws quickly gained a reputation for corruption and haplessness. Until the revenue service was professionalized in the mid-nineteenth century, an individual did not have to possess any knowledge of Ireland or distillation to become an excise officer. Many excise officers extracted bribes from distillers. Some did this because they were corrupt, others because the government paid such paltry salaries. Asset forfeiture policies authorized revenue officers to seize any materials used to produce and traffic illegal whiskey, including stills, wagons and horses. Brawls erupted between citizens and excise officers, and at times England's soldiers were sent into Ireland to aid in the war on poteen.

Irish Whiskey Booms

Sensibly, the English government upgraded its whiskey policies. The Excise Act of 1823, which applied to Ireland and Scotland alike, allowed a whiskey-maker to register his still with the government with little trouble. He would be taxed only on the quantity of whiskey he produced. The process of professionalizing the whiskey regulation and taxation process had begun, and over the next century excise officers and distillers slowly developed a professional working relationship. Better laws and better enforcement encouraged more compliance and reduced the incentives to make poteen.

The timing of this change in policy was fortuitous—the industrial revolution had begun in whiskey. Tinkerers had been devising machines that could turn whiskey-making from a crude, labour-intensive process into an efficient, mechanized system of production. While some profits could be made producing illegal poteen, the big money could be had by setting up a whiskey factory legally, and taking advantage of economies of scale in both grain purchases and mechanized output.

Irish whiskey rapidly became big whiskey, outpacing Scotch. Midleton distillery, for example, set up a colossal 31,500-gallon pot still in 1825. Dublin, meanwhile, grew into a powerhouse of whiskey production. The companies of John Jameson, William Jameson, John Power and George Roe grew into hulking firms. Roe's Thomas Street plant spanned 17 acres and churned out 2 million US gallons (7.5 million litres) of whiskey per year. Production of legal whiskey quadrupled over the next 75 years.

One mid-eighteenth century visitor to Ireland reported that in one small town nearly every public house sold whiskey, which locals esteemed 'as a wholesome balsamic diuretic; they take it here in common before their Meals. To make it more agreeable they fill an iron pot with spirit, putting sugar, mint and butter.' After heating it a while, they poured it into cans, toasted each other, and drank it down.

Ireland's whiskey also was consumed outside the nation in the world's finest eateries and clubs. Australia, India and the United States also imported Irish whiskey. The Irish Whiskey made by Coleraine in the north (County Derry) was served in the UK Parliament's House of Commons.

Too Much of a Good Thing

The growth of the Irish whiskey industry was not without its costs, of which the most obvious are alcoholism and the health and social ills that go with it. Prior to the twentieth century, reliable data on the rates of alcoholism in Ireland are nonexistent. However, many of the writings of the day portray the ugly effects of too much whiskey. Unsettlingly, the themes of whiskey abuse and senseless violence are often intertwined in these depictions.

The great wit Jonathan Swift (1667–1745) described an Irish feast where pails of 'usquebaugh' were served. By the night's end, things turn ugly for no obvious reason.

> What stabs and what cuts,
> What clattering of sticks,
> What strokes on the guts,
> What bastings and kicks!
> With cudgels of oak,

> Well hardened in flame,
> An hundred heads broke,
> An hundred struck lame.

A tune by the Irish composer Turlough Carolan (1670–1738), a contemporary of Swift's, exhibits one man's unhealthy relationship to whiskey:

> That you may be swallowed, oh, whiskey!
> It's you've left my windpipe scalded;
> It's often you left me empty
> Without silver or gold.

Addicted, this hapless drunk fails to attend Mass, skips meals in favour of boozing and loses his good reputation. Yet, he continues:

> Sure you dirty my clothes
> And sure you cut my nose.
> You knock me to the ground,
> And you leave me without energy;
> But you cure me in the morning,
> And I forgive you for the future.

This lyric might have been autobiographical. The writer Oliver Goldsmith (1730–1774) said that Carolan 'would drink whole pints of usquebaugh', and that this habit 'brought on an incurable disorder' that killed him.

William Carleton's *Traits and Stories of the Irish Peasantry* (1843) describes the weird joy of the men who partook in the brutal gang fights between Catholic 'Ribbonmen' and Protestant 'Orangemen'. For them, whiskey, was an 'ardent spirit' that stoked the 'hot blood' of nationalism. Jaws were shattered, necks and bones were broken, all of which had the odd effect of making the hopped-up fighter feel 'in love with everyone, man, woman, and child'.

Of course, as so often happens, the Irish did find some humour in the awfulness, perhaps most famously in the ballad 'Finnegan's Wake'. Tim Finnegan drinks whiskey each morning. One morning, he suffers delirium tremens atop a ladder and falls and breaks his skull. At his alcohol-soaked wake, a brawl breaks out and Tim is revived from the dead when uisce beatha, the water of life, is slopped on his corpse.

Similarly, the anonymously authored nineteenth-century tune 'A Sup of Good Whiskey' warns of whiskey abuse but then tweaks society's do-gooders.

> A sup of good whiskey will make you glad;
> Too much of the creatur' will make you mad;
> If you take in reason, 'twill make you wise;
> If you drink to excess, it will close up your eyes…
> Some doctors will tell you 'twill hurt your health;
> The justice will say 'twill reduce your wealth;
> Physicians and lawyers both do agree,
> When your money's all gone, they can get no fee.
> Yet surgeon and doctor,

And lawyer and proctor,

Will all take a sup in their turn.

Irish Whiskey Busts

Irish whiskey peaked in 1900. Thirty distilleries in the nation produced a record 9.9 million US gallons (37.5 million litres) of whiskey that year. Irish whiskey was renowned for its quality. Twenty years later, the industry was wrecked. What happened? In short, many bad things, some avoidable, others inescapable.

Beginning in the 1840s, a social backlash against drunkenness—the temperance movement—began. Almost concurrently, the country lost nearly one quarter of its people (and drinkers) to the death and migration that ensued during the Great Famine. Both of these factors limited the long-term growth of the home market for Irish whiskey.

Despite this, Irish whiskey production grew over the century, thanks to demand in England and overseas. Between 1870 and 1900, Irish whiskey production doubled, and between 25 and 60 per cent of Irish whiskey was exported.

Unfortunately for Ireland, it was not alone in the whiskey market. As the previous chapter described, the Scottish distilling industry had been expanding since the 1830s, fuelled through the use of the patent still. Most of the big Irish distillers based in Dublin and Midleton disdained the patent still for producing inferior whiskey. In Northern

Ireland and Scotland, though, the patent still was more quickly adopted by whiskey-makers.

The barons of Irish whiskey complained to the UK government, and repeatedly asked it to require whiskey and blended whiskey to be labelled differently. True Irish whiskey, they contended, was made from malted barley in a pot still. In 1890, after much squabbling, the Select Committee on British and Foreign Spirits could only agree that whiskey was 'a spirit consisting of alcohol and water'. This left patent still operators free to continue churning out grain whiskey and labelling it 'whiskey'.

The Irish distillers' decision largely to spurn the new technology was costly, and advantaged their competitors. Both the Scots and producers in Northern Ireland used the patent still to grab positions in the growing English re-distillation market. (In the nineteenth century, much of the whiskey produced in Ireland and Scotland was shipped to England to be re-distilled and made into gin.) The patent still enabled both Scotland's and Northern Ireland's distillers to produce low-priced spirit for this market. Additionally, these whiskey-makers gained traction in the consumer market. They and the whiskey dealers they did business with mixed patent still whiskey with more robustly flavoured Irish pot still whiskey or Scottish pot still whiskies to produce low-priced, easy-sipping blended whiskies. Consumers in England and abroad quickly took to them.

Religion and politics came to taint the dispute over what realwhiskey was. Much of the blended Irish whiskey was produced in Northern Ireland. Greedy Northern Ireland producers were blamed for bastardizing Irish whiskey and ruining the industry. The whiskey in one's glass came to be viewed as a political statement. Buy a glass of Dublin-produced Jameson's or Power's and you were supporting the nascent Irish state; order a glass of Bushmills and you were supporting England.

The bad times began in 1901. Over-production of spirit combined with a general economic downturn in England and Europe left all whiskey-makers with more booze than they could sell. Production slid, and prices were slashed. Between 1900 and 1915, Irish whiskey production fell 29 per cent, from 9.9 to 7 million US gallons (37.5 to 26.5 million litres). The number of Irish distilleries atrophied from 30 to 21.

The First World War wreaked further havoc on the Irish whiskey industry. The markets for all alcoholic beverages shrank violently, and the Chancellor of the Exchequer, David Lloyd George, used the war to pursue prohibitionist policies. When stories of industrial workers being too drunk to show up to work circulated, George and fellow social reformers seized upon these anecdotes and claimed that these sots were undermining the war effort.

Despite the fact that nearly 15 per cent of the UK's tax revenues were provided by the spirits industry, and 30 per cent by alcoholic beverages generally, Parliament established the Control Board (Liquor Traffic) in 1915. This body was empowered to take measures to curb alcohol consumption in areas where war materiel was being produced or transported.

The Board interpreted its mandate very broadly, and soon imposed restrictions of the sales and consumption of alcohol in England, Wales and Scotland, and eventually in five-sixths of the UK. In the name of 'principles of physiology', the 'principles of hygiene' and the government goal of 'national efficiency', the Board curbed the public's drinking. The only areas spared the Board's rod were remote rural patches of the kingdom. Add to these onerous measures the near-quintupling of the tax on spirits and the clamp-downs on production, and it is not difficult to see why so many distilleries went out of business.

Unlike the Scottish whisky-makers, who also suffered, the Irish distillers lacked sufficient financial strength and overseas sales to survive these blows. Regrettably, the Irish distillers' tenacious cling to their pot stills and barley left them unable to take advantage of the UK government's demand for massive quantities of industrial alcohol. The Irish whiskey-makers were dealt a near death blow when the UK government issued the Barley (Restriction) Order in 1917. This wartime measure forbade the use of barley for anything but producing food. Thanks to their adoption of the patent still, many Scottish whisky-makers survived. In 1915 Scottish whisky production had jumped back to 18.5 million US gallons (70 million litres), almost twice that of Ireland at its peak. The Scots suffered less from the barley policy since they grew and imported corn and other cereal grains for feeding into their stills.

Ireland and England's inability to cooperate politically

inflicted a near mortal wound on Irish distillers. Ireland's rebellion against England, which began in 1916, resulted in the establishment of an Irish state separate from Northern Ireland. Unfortunately, this win for Ireland sparked trade wars, which kept Irish whiskey from English mouths.

The sales of Irish whiskey took an additional hit when the US government enacted Prohibition in 1920. Whiskey imports to America were forbidden, and a huge Irish whiskey market was lost for over a decade. More distilleries closed or merged their operations to avoid ruin, and production slid further.

When Prohibition ceased and the US market re-opened in 1933, Irish distillers lacked sufficient whiskey stocks and production capabilities to take advantage of the surge in demand. Even more dreadful was that much of the Irish whiskey that did exist could not be sold to the United States. A 1926 Irish law had raised the bonding (ageing) period for Irish whiskey from three to five years. The notion was laudable—make Irish whiskey a super premium product by requiring it to age much longer than other types of whiskey. But the short-term effects were horrendous. The US had a law that forbade the sale of imported booze that could not be sold in the nation that produced it. This meant that Irish whiskey less than five years of age could be sold in neither Ireland nor America.

Out of the Ashes

For decades, the Irish whiskey industry shrivelled, its annual production during the first half of the century was perhaps one-tenth of what it was in 1900. The Irish government, in its grabs for much-needed revenue, raised the taxes on whiskey again and again. Between 1900 and 1969, the duties were increased 2,600 per cent.

Nevertheless, things began to look up around mid-century. Helpfully, the Irish government chose to get behind Irish whiskey as a product. In 1950 it established legal definitions of 'Irish whiskey' and 'Irish Pot Still Whiskey'. Both had to be made in Ireland with malted barley. The latter had to be wholly distilled in copper pot stills, the former could include patent still spirit. The government also began to help fund advertising campaigns that sold Irish whiskey to overseas consumers. In 1964 the government enacted a tariff on distilled spirits produced outside Ireland, which made Irish whiskey more attractive to domestic consumers. Five years later, it reduced the time that Irish whiskey had to stay in barrel from five to three years, which helped distillers to sell their whiskey more quickly and with less evaporation.

Irish whiskey-makers also wised up. They adopted the patent still and diversified their products, churning out low-cost Irish vodkas and gins. The Tullamore distillery produced Irish Mist liqueur, a concoction of Irish and grain whiskies, honey and herbs. It marketed it as an ancient drink first consumed by the Celts a millennium ago. Irish Mist became fantastically popular in the US and Europe. Though the Tullamore distillery did not survive, Irish Mist lives on, as does the distillery's famed brand, Tullamore Dew. Irish whiskey sales got a further lift in the 1940s and 1950s when the idea of mixing Irish whiskey with coffee, sugar and cream to produce Irish coffee spread around the globe.

To take advantage of economies of scale and reduce competition between Irish whiskey distillers, the few remaining firms merged to form United Distillers of Ireland in 1966. (Two years later, the firm renamed itself the Irish Distillers Group Ltd, or IDGL.) All the former competitors, John Jameson & Son, the Cork Distilleries Company and John Power & Son, became brethren brands. Come 1975, all these brands were produced at a new, state-of-the-art facility in Midleton. IDGL ceased sales to firms that blended Irish whiskies and sold their whiskies under their own brand names. IDGL coordinated the production and marketing of all its brands, targeting each brand to particular markets. It bet its future on blended Irish whiskies, not pure pot still whiskies.

Meanwhile, in Northern Ireland, the distilling industry nearly collapsed. It could not keep pace with the Scottish distillers, and demand in its home market was weak. In 1947 the Bushmills and Coleraine distilleries partnered. Old Comber shut down shortly thereafter. Coleraine was absorbed by Bushmills in 1964, and its stills were shut off in the late 1970s.

The 1970s marked both an end and an ironic new

beginning for Irish whiskey. Since their earliest days, most Irish distilleries had been run by Irishmen. In many instances, leadership of the firms was passed down from generation to generation, or shifted to friends of the family whose stewardship could be trusted.

That ceased in the 1970s. Bushmills was bought up by the UK's Bass-Charrington, the owner of many pubs. Bushmills got a sales lift; its oblong bottle now appeared on bar shelves all about the kingdom. Then in 1972, the once unthinkable occurred—the old Ireland-Northern Ireland whiskey rift closed. Seagram, the Canadian-based booze colossus, pur- chased Bushmills, then swapped it to IDGL in exchange for 15 per cent of IDGL. All Irish whiskies were in it together.

Irish Whiskey at the Close of the Twentieth Century

As the century wound down, Irish whiskey got better and better. IDGL re-engineered Jameson, the old Dublin brand, turning it into a light, smooth and fruity Irish blend. It adroitly packaged both the newJameson and Bushmills and grew their sales steadily over the next fifteen years. Gone was the 'shamrocks and shillelaghs' approach to selling Irish whiskey. Brands were repositioned as upscale and sophisticated spirits consumed by cosmopolitan people about the world.

In 1988 IDGL fended off a British company eager to snatch it up by selling itself to Pernod Ricard, a French megacorporation that owned a diverse portfolio of beverage brands, including Beefeater Gin and Yoo-hoo, a non-alcoholic chocolate milk drink. Pernod Ricard renamed the group Irish Distillers Ltd, and grew its whiskey sales further. In 2007 sales of Jameson rose 15 per cent to over 2.6 million cases (about 6.5 million US gallons or 25 million litres), with hefty sales growth over the past year in Bulgaria (+61 per cent), Russia (+41 per cent), the Ukraine (+43 per cent) and the US (+21 per cent).

Meanwhile, in 2005 the Bushmills distillery and brand was sold to the London-based Diageo, the owner of Smirnoff vodka, Tanqueray gin and Ireland's Guinness beer and Baileys liqueur, for $324 million. Diageo is attempting to grow Bushmills, and it reported double-digit sales growth in eastern Europe in 2008. It intends to increase annual production from 950,000 US gallons to 2.4 million (3.6 million to 9 million litres) come 2011. These days, Bushmills is offering more versions of itself, such as Blackbush, White Bush and ten- and sixteen-year-old Bushmills.

Under this Irish whiskey revival, consumers now have more and better choices. Twenty years ago, only a couple of brands were available outside Ireland. Now, it is not unusual to have a half dozen or more Irish whiskies to choose from—including pure pot still whiskies like Redbreast, single malt whiskies like Locke's, and super-luxury whiskies such as Midleton Very Rare, which costs over £120 or $100 a bottle.

In an additional sign that the rebirth of Irish whiskey is underway, the two hulks of Irish whiskey have been joined by an uppity peon: the Cooley Distillery, which is located in County Louth. In its mere two decades, Cooley has landed an armload of medals and awards for its whiskies, which include Kilbeggan, the Tyrconnell, the peated Connemara and others. Cooley delivered 690,000 gallons (2.6 million litres) of whiskey to 40 nations in 2008, and it is increasing its production further.

Collectively, the Irish distillers now are selling 9.5 million US gallons (36 million litres) of whiskey per year—which is not much in comparison to the 215 million US gallons (814 million litres) of Scotch, but is still a huge improvement.

Chapter 5
Whiskey in the USA

In Scotland and Ireland, whiskey-makers were greatly affected by the actions of the English government. Its policies made whiskey both worse and better. In the USA, distillers faced no such trouble. British rule ended in 1782, and the new US government rarely intervened in booze-making. This laissez-faire environment encouraged a whiskey boom, but the boom awakened a harsh social backlash that culminated in a government ban on alcohol. Eventually, the nation made peace with whiskey and it was up, up and away as American whiskey quickly went global.

Early History

The United States of America came into existence in 1789, but long before that European settlers were distilling whiskey. While authoritative public records are limited, plenty of other scraps of evidence exist. For example, a Virginia farmer, George Thorpe, wrote to a relation in England in 1620: 'Wee have found a waie to make soe good drink of

Indian corne I have divers times refused to drinke good strong English beare [beer] and chose to drinke that.' By 1645, Virginia's government had set a price cap of £40 per gallon for brandy and 'aqua vitae'.

Four hundred miles northwards in what would become New York City, the Dutch West India colony had a distillery that may have made hooch from corn and rye during the 1640s. And 300 miles to the south of Virginia, Thomas Ashe, a writer who visited the area that would become South Carolina, reported in his *Carolina or a Present Description of the State of That Country* (1682) that the settlers drank their corn. They had 'invented a way of with it good sound Beer: but its strong and heady: By maceration, when duly fermented, a strong spirit like Brandy may be drawn off from it, by the help of an Alembick.'

As in Ireland and Scotland, some local political leaders worried that farmers were diverting too much grain from bread production to whiskey-making, and so issued laws and court orders to temporarily curb or suspend distilling. For example, in 1676 a court in Pennsylvania allowed locals to distil only grain 'unfit to grind and boalt'. And officials were also unnerved by the effect of whiskey and liquor generally on public morals. Ministers and religious leaders were taken to task for setting a bad example. In 1631 Virginia's leaders declared that 'mynisters shall not give themselves to excesse in drinkinge, or riott'.

Though whiskey was much made in colonial America, it was not the most popular liquor. Fruit brandies were widely consumed, but rum was the undisputed king. The infamous 'triangular trade' had America shipping rum to Africa, Africa shipping slaves to the Caribbean islands to work sugar plantations and these islands shipping sugar and molasses to America for distillation into rum.

Yet rum's run as America's top booze ended quickly and whiskey took its place. The early USA was an agrarian nation flung in small settlements over a widespread area; corn, rye and wheat grew well in much of the country. Farmers harvested grain sufficient to feed their families, and sold the rest. Grain might or might not fetch a fair price at market, but liquor always sold well. Hence farmers and millers had a strong incentive to produce whiskey, and produce they did. Meanwhile, the young nation's development of roads and navigable waterways made it easier to bring booze to distant markets. Production soared, and prices fell. The Port of New Orleans saw 11,000 US gallons (41,600 litres) of whiskey arrive in 1812; four years later, it handled 320,000 US gallons (1.2 million litres).

The rum trade, meanwhile, was disrupted by the waning of the international slave trade (both the USA and Britain began getting out of the business at the start of the nine-teenth century), and the trade wars—and armed wars (1775 to 1783 and 1812 to 1815)—between the USA and Britain. Additionally, US settlers began to associate rum with Britain and its much-loathed Royal Navy. Domestic boosters touted whiskey as a native spirit, and consumers happily paid less for it than for rum (which was heavily taxed). By 1810 US whiskey consumption had exceeded rum drinking.

Emblematic of whiskey's ascendancy as America's preferred spirit was that George Washington, the nation's first president and 'founding father', was a whiskey man. While commanding troops during the American Revolutionary War, General George Washington pleaded with his civilian commanders for whiskey. '[T]there should always be Sufficient Quantities of Spirits with the Army… In many instances, such as when they are marching in hot or cold weather, in Camp or Wet, on fatigue or on Working Parties, it is so essential that it is not to be dispensed with.' Washington suggested that the government erect a whiskey distillery in each of the states to supply the troops. His proposal went nowhere.

After serving as president (1789–97), Washington retired to his Mount Vernon farm. James Anderson, a Scotsman in his employ, urged him to build a distillery. Washington did, and within a year his distillery was producing 11,000 US gallons (41,600 litres) of rye and corn whiskies. At Anderson's suggestion, Washington raised hogs, fattening them on the used grain from the mashtun.

Whiskey Booms

The nineteenth century was a boom time for whiskey in America. The population grew from 5.2 million to 76.2 million; more people and more farms meant more whiskey.

Technological innovations also helped—dozens of new distilling technologies, such as steam coils to heat stills, were patented by profit-hungry inventors.

Additionally, US tax policy boosted whiskey. In short, the US government imposed excise on booze imports, but rarely on domestic liquor. The young nation's government had learned a lesson in 1791 when it tried taxing home-produced spirits. Distillers and drinkers attacked tax collectors and an armed insurrection (The Whiskey Rebellion) broke out the next year. President George Washington had to send nearly 13,000 federal troops to quash it. The first permanent national tax on whiskey did not come until 1862. As in the UK, the justification for the excise was war—the Civil War.

Generally, the US government treated the industry as a partner, rather than a sheep to be fleeced. Whiskey taxes generally were kept low; sensibly, the federal government permitted distillers to avoid paying taxes on whiskey while it aged in warehouses. This 'bonding' practice, which began in 1868, was gradually extended to permit distillers to hold their whiskey for up to twenty years before having to pay excise. This win-win practice encouraged whiskey-makers to age their product, thereby improving its taste. The Congress also exempted distillers from having to pay tax on the whiskey that evaporated during ageing (1880) and it passed a 'bottled in bond' labelling law (1897). This was a mark of quality, insofar as it indicated that the whiskey within a bottle was made in accordance with exacting government regulations. The federal government, meanwhile, saw its coffers swell.

Whiskey taxes provided half the government's revenues.

As in Ireland and Scotland, the continuous still created chaos in the whiskey industry, as it raised the age-old question, 'What is whiskey?' No law prohibited the mixing of a small amount of high-quality whiskey with cheap, neutral grain spirit, darkening its colour with prune juice and peddling it as whiskey.

The Pure Food and Drug Act of 1906 helped deal with this problem. It provided the federal government with the authority to require consumable products to be safe and properly labelled. President Theodore Roosevelt's administration (1901–9) issued regulations to require whiskey to be labelled as 'blended whisky', 'compounded whisky' or 'imitation whisky'. Since then, these labels and definitions have changed, but the principle remains: do not deceive the customer—call the hooch what it is.

Too Much of a Good Thing

Regrettably, some Americans began to abuse whiskey almost as soon as it appeared. The 1671 *Minutes of the Executive Council of the Province of New York* include this recommendation of an official from another province:

> That ye distilling of Strong Liquor out of Corne, being ye Cause of a great Consumption of that Graine, as also of ye Debauchery & Idleness of ye Inhabitants, from

whence inevitably will follow their Poverty and Ruine,
bee absolutely prohibited or restrayned.

At this time, some colonies already were enacting bans on the practice of paying American Indians with liquor. But these bans were to little effect. An excess of cheap liquor made from corn was available nearly everywhere, and settlers continued to use it as money. Alcoholism ravaged the Native American population. In 1754, as Hugh Lefler reports in *North Carolina History Told by Contemporaries* (1934), a leader of the Catawbas issued a plea to the Americans:

brothers here is One thing You yourselves are to Blame very much in, That is you rot your grains in Tubs, out of which you take and make Strong Spirits. You sell it to our young men and give it to them... [T]he effect of that drink is very bad for our people, for it Rots their guts and Causes our men to get very sick.

In the nineteenth century, the US became an 'alcoholic republic', as historian William J. Rorabaugh put it. In 1830 the average man and woman over the age of fifteen was consuming 9.5 US gallons (36 litres) of liquor per year. (Today, Americans consume a little over 0.7 gallons (2.65 litres) of spirits each year.) Cheap whiskey was available everywhere, and Americans drank lots of it. Whiskey in the morning, whiskey whenever a guest came to visit, whiskey whenever a child was born or someone died. Whiskey was administered as a medicine and used in bizarre 'cures'. A remedy for dropsy required the afflicted patient to chug 'day and night' a mixture of whiskey, mustard seed, beets, horseradish and brown eggshells. If health did not return, the remedy advised the invalid to 'repeat the same'. Anne Royall, a nineteenth-century American journalist who toured the eastern parts of mid-America, was appalled at what she saw. 'When I was in Virginia, it was too much whiskey—in Ohio, too much whiskey—in Tennessee, it is too, too much whiskey!'

Whiskey also played an unseemly role in elections. George Washington, when running for office in Virginia in 1758, supplied voters with plenty of drink and won office. James Madison, who later became America's fourth president, lost a 1777 state election because he refused to buy votes with whiskey. Beyond providing voters with a free binge, whiskey giveaways were viewed as a sign that the candidate cared about the common man.

The excess consumption resulted, as Rorabaugh explains, in 'wife beating, family desertion, and assaults, as well as payments from public funds for the support of inebriates and their families'. Some made light of this insanity. Kentucky resident Thomas Johnson, Jr (1760– 1820), often referred to as the 'drunken poet of Danville', composed this doggerel for his epitaph:

> Underneath this marble tomb,
> In endless shade lies drunken Tom;
> Here safely moor'd, dead as a log,

Who got his death by drinking grog,

By whisky grog he lost his breath

Who would not die so sweet a death.

Much of the USA, which is moralistic by nature, was unnerved at the rising social disorder. Temperance societies that ballyhooed God and condemned grog were rapidly formed. As more and more Americans flocked to these organizations, booze consumption dropped. Temperance groups, nevertheless, further ramped up their anti-alcohol campaigns. They circulated tens of millions of pamphlets preaching abstinence and telling often ludicrous tales of the horrors of alcohol. One story had a drunkard losing his leg in an industrial accident, then selling the severed limb to a disreputable surgeon, and using the proceeds to go on a bender. Other pamphlets tapped into American ideals of freedom and self-reliant independence. Drinking was likened to the 'servile' life under King George or to slavery, and tipplers were cajoled to cast off their chains by putting down their cups. Drunks were damned for failing to be economically productive citizens.

The cause of drinkers (wets, as they became known) was further hurt by the infamous Whiskey Ring Scandal of 1875. Big distillers had bribed government officials to dodge paying millions in federal alcohol taxes. President Ulysses S. Grant's personal secretary was indicted. A media-political circus ensued.

Finally, some distillers and sellers further sullied

whiskey's name by hawking it as a medicinal cure-all. The 14 June 1900 edition of the *Evening Times* of Washington, DC carried a large advertisement with the lead 'Four Million Cures—No Failures'. Duffy's Pure Malt Whiskey, it said, had cured 164,326 cases of dysentery, 331,521 cases of malaria and 331,246 cases of 'weak women'. Health was a sip away—'If you are half sick it is because your blood is out of order. You need a stimulant. Take DUFFY PURE MALT WHISKEY as directed. You will be cured and your system will not be injured as with deadly drugs.'

The die was cast—whiskey was associated with immorality, shady dealings and vice.

Prohibition: An American Folly

During the First World War, liberty-loving Americans acceded to numerous government demands for sacrifice for the good of the nation. Free speech was curtailed, and agricultural and industrial output was directed and rationed by the government. Consumable alcohol production plunged, as distilleries were used to make industrial alcohols for war-making.

As the 'war to end all wars' came to an end, the temperance groups made a manoeuvre that likely had Lloyd George applauding—it aggressively lobbied the government to turn the war-time booze-control measures into an outright ban on alcohol.

For years, distillers and drinkers had been fighting back. They organized anti-anti-alcohol rallies, and took to the newspapers to denounce the encroachment on their liberties by a zealous minority. George Garvin Brown (1846–1917), who founded the Louisville, Kentucky whiskey company that grew into the mega-distiller, Brown-Forman Corporation, penned *The Holy Bible Repudiates Prohibition* (1910). In it, he cited the Bible's apparent support for alcohol beverage consumption, and declared, '[T]here is no more moral turpitude in manufacturing and selling an intoxicating liquor than there is in manufacturing and selling any other product... [M]an is responsible to God for his every act.'

It was to no avail; the 'drys', as they were called, swamped both federal and local politicians with letters and petitions and convinced them to enact the Volstead Act (1918) and the Eighteenth Amendment to the US Constitution (1919). Alcoholic beverages were criminalized, and both the national and state governments were empowered to loose armed officers of the law on everyone who manufactured or trafficked in them.

While Prohibition made legal liquor less accessible, there were legitimate ways to score a bottle of whiskey. The most obvious loophole was the 'doctor's orders' excuse. The law read,

> No one shall manufacture, sell, purchase, transport, or prescribe any liquor without first obtaining a permit... except that a person may, without a permit, purchase and

use liquor for medicinal purposes when prescribed by a physician... Not more than a pint of spiritous liquor to be taken internally shall be prescribed for use by [a sick] person within any period of ten days and no prescription shall be filled more than once.

The American writer Harry Kroll (1882–1967) saw the law's loophole and acquired bottles with the help of his physician, who prescribed alcohol to him for 'colds, flu, muscular aches and pains, constipation, diarrhea, pin worms, and piles'. Tongue planted in cheek, Kroll noted, 'Once I used [whiskey] for jock strap itch, and discovered if taken internally the effect was more lasting and much more palatable.' To supply Kroll and other tipplers, the government licensed the American Medicinal Liquor Company, a consortium of six Kentucky distillers. It churned out an eye-popping 1.4 million US gallons (5.3 million litres) of bourbon each year.

Prohibition was difficult to enforce. People devised ingenious ways to hide whiskey. Government officials found that oblong bottles of Johnnie Walker Red Label were being smuggled inside loaves of bread. Additionally, the continental US is an enormous land mass—over 3 million square miles. Stopping everyone everywhere from producing alcoholic beverages was an administrative impossibility—it would require an army of millions of officials with nearly boundless powers to search and seize.

Warren G. Harding, who served as US president from

1921 to 1923, made a mockery of Prohibition. An avid golfer, Harding consumed whiskey cocktails on the links. Many others in the nation's capital disobeyed. Bootleggers brought rye whiskey in from Maryland, and bourbon flowed behind closed doors in the Capitol building, which holds the US Congress. Speakeasies flourished in New York City, where smugglers could bring Irish and Scotch whiskey via small boats or hidden in cargo ships carrying other items. Drinkers who wished to avoid prosecution boarded cruise ships that sailed from the city into international waters, where whiskey and other boozy beverages flowed legally.

Those in America's northern states crossed the border to get their spot of whiskey. This bit of doggerel of the day captures their less than reverent attitude towards federal authority:

> Four and twenty Yankees feeling rather dry;
> Marched into Canada to get a little rye.
> When the rye was opened the Yanks began to sing;
> Who in the hell is Coolidge, and God save the king.

Coolidge, of course, would be Harding's successor, President Calvin Coolidge, who tried to make Prohibition work. Canadian whisky flowed into America. According to a February 1930 Associated Press report:

> Large quantities of Canadian beer and whisky are being transported in cars from Amherstburg, Ontario,

Canada, across the frozen lower Detroit River, to the
Michigan [US] side of the international boundary line...
The cars are driven with one door open, so if the car goes
through the ice the driver can scramble free.

Whisky had been made in Canada since the 1700s.
Many of its early settlers were from Scotland, and not
long after they put down roots in the areas near Lake
Erie, Lake Ontario and Toronto, grain mills and attached
distilleries appeared. As in other nations, Canada's distillers
industrialized in the nineteenth century, building huge whisky
factories, like that of Gooderham and Worts on Toronto's
waterfront. The US's experiment with Prohibition had at
least one positive effect—it spurred growth in the Canadian
whisky industry. Canadian whisky poured over the nations'
enormous shared border. Whisky-makers like Hiram Walker,
Joseph Seagram and Samuel Bronfman became industry
titans as Canadian whisky went gangbusters. Canadian Club
became a mega-brand, and it and other Canadian whiskies
were advertised in *Life*, *Playboy* and other major US magazines
until the 1970s.

Beyond its inherent silliness, Prohibition also had
pernicious effects. The US government lost billions in tax
revenues when legal sales of alcohol plummeted, and droves
of breweries and distilleries went out of business. Meanwhile,
vicious criminals got rich making and smuggling booze.
Perhaps 45,000 Americans died from drinking poisonous
spirits produced by unscrupulous criminal gangs, while

others suffered permanent neurological damage.

Mercifully, the US government ended its peculiar experiment in anti-alcohol policy on 5 December 1933. Alcohol-lovers blasted off cannons, and flooded the streets and drinking establishments across the country. Tax revenues poured into the US Treasury and workers returned to the payrolls of beverage producers, two positive developments for a nation mired in the Great Depression.

Yet, even today vestiges of America's split-mindedness over booze remains. It is exemplified by the paradox of the Jack Daniel's distillery in Lynchburg, Tennessee. Hundreds of thousands of visitors make their way to the famed distillery, which sold more than 2.2 million US gallons (8.3 million litres) of 'JD' in 2008. Yet none of these whiskey pilgrims can buy a bottle there. The distillery is sited in Moore County, a 'dry county' where alcohol sales are forbidden. Moore County is one of approximately 500 dry localities in the United States. Meanwhile, fourteen of 50 states still ban liquor sales on 'the Sabbath' (Sunday).

And whiskey continues to be singled out for afflictions suffered by alcoholics generally—whiskey dick (impotence), whiskey shits (diarrhoea) and whiskey face (broken facial blood vessels). One does not hear of vodka shakes or wine dementia, and the only pejorative associated with beer is 'beer gut'.

American folk, blues and country music long have reinforced the image of whiskey as a bad thing. Amos Millburn's popular 1950 tune, 'Bad, Bad Whiskey', describes a man who tries to stay sober but succumbs to whiskey and loses his home. Similarly, in 'Don't Sell Daddy Any More Whiskey' (1954), a family goes hungry because its father spends its money on whiskey binges. At the song's conclusion, a baby wails in misery. Other US whiskey-abuse songs are confessionals: the singer admits that he is getting dead drunk alone, usually over love lost. John Lee Hooker's 'One Bourbon, One Scotch, One Beer' (1966) and Hank Williams, Jr's 'Whiskey Bent and Hell Bound' (1979) are of this ilk. Some famously raucous US artists have taken the badness of whiskey and made it cool, an affect of their 'renegade' lifestyles. Singer Janis Joplin (1943–1970) often had two stools on stage; she sat on one, and a bottle of Southern Comfort sat on the other. As a member of the rock group Van Halen, Michael Anthony (1954–) played a bass guitar shaped like a Jack Daniel's bottle and sometimes chugged JD straight from the bottle. The writer Hunter S. Thompson (1937–2005) frequently portrayed himself as consuming whole bottles of Wild Turkey bourbon, running amok at events like the Kentucky Derby horse race, and driving while clutching a tumbler of whiskey.

A Rough Century Concluded with a Boom

The end of Prohibition did not bring the onset of happy days for the US whiskey industry. Half the distilleries in Kentucky were lost for ever, and few of the nation's small whiskey-making companies returned to operation. The shrunken, feeble industry had only a decade to produce before the US entered the Second World War, and distilleries had to turn to making industrial alcohols that could be used to produce mortars and other war materiel.

Additionally, Prohibition had shifted US drinkers' tastes towards lightly flavoured spirits, like gin and vodka, which bootleggers could churn out with ease. What whiskey was to be found often came from Canada, and it too was very light-bodied, being a mixture of whiskey and neutral spirits. Or it came from Scotland. This left cash-hungry, US whiskey-makers playing catch-up. To regain market share, they had a choice: either they could ape Scotch whisky (a near impossibility) or they could churn out young, bland, Canadian-style whiskies. They chose the latter course, which all but spelled the end for rye whiskey, a bold spirit that shows peppery, rich aromas and tastes that had been made since the Colonial era.

In retrospect, American distillers' decision to play follow-the-leader and shift away from bolder flavoured whiskies is understandable, but as a strategy it was questionable. They abandoned what they did well, and Americans who craved light spirits kept buying Canadian whisky, much of which they swilled in mixed drinks, like the 7-7 (Seagram's Seven and 7UP). Or they purchased vodka, which companies marketed as a 'clean' and 'pure' liquor that did not leave one's breath smelling boozy. Consumers who craved more flavourful booze purchased blended Scotches, which were marketed as high-class imports that gave drinkers status.

Things began to turn around for American whiskey in the 1980s. Since Julia Child had taken to trying to teach Americans to 'master the art of French cooking', a quality food and drink movement had taken hold. People began questioning why so much of their food was pre-made and came from tins, and they were entranced at the prospect of tastier and healthier fare. California vintners, long known for making cheap jug plonk, started producing world-class wines; home and craft brewers took on Miller and Anheuser-Busch with their small production brews which oozed malt and hops.

US distillers were slow to catch on, but when they did the time was right. The alcoholic and drug excesses of the 1960s and '70s had left the nation blasé. Individual consumption of liquor had grown from 0.7 US gallons (2.65 litres) per year to 1.07 gallons (4.05 litres) between 1948 and 1978. (In 2006 it was 0.71 gallons or 2.69 litres.) In possession of more disposable income than ever, Americans were amenable to the idea of 'drinking less, but drinking better'.

Due to industry consolidation, much of American whiskey production had shifted to the states of Kentucky and Tennessee. (Just before Prohibition, Illinois and Indiana produced more booze than Kentucky. Sixty years later, Kentucky made four times as much whiskey as Indiana, and fifteen times as much as Illinois.) Distillers put their energies behind producing two styles of whiskey—Tennessee whiskey and bourbon.

The former is produced by just two companies, the small but high-quality George Dickel Distillery, and the BrownForman Goliath, the Jack Daniel's Distillery. Tennessee whiskey, as chapter One mentioned, differs from bourbon in being filtered through charcoal prior to barrel ageing. Sales of Jack Daniel's grew 65 per cent between 1998 and 2007.

Bourbon, meanwhile, is made by many companies, nearly all of which are in Kentucky. Much to the dismay of rye and Tennessee whiskey drinkers, bourbon whiskey is increasingly being deemed America's spirit. In part, this is entirely sensible—distilling corn into spirit and ageing it in charred barrels originated in the US. Obviously, the bourbon industry advocates this view, and bourbon's advocates like to point out that in 1964, Congress enacted a resolution declaring bourbon to be 'a distinctive product of the United States... [that has] received recognition and acceptance throughout the world'. This is true, but only partially—the whole of the resolution is not so much a christening of bourbon as America's official hooch; rather, it is a demand for trade protection. The resolution concludes:

> That it is the sense of Congress that the recognition of bourbon whiskey as a distinctive product of the United States be brought to the attention of the appropriate agencies of the United States government toward the

end that such agencies will take appropriate action to prohibit the importation into the United States of whiskey designated as 'bourbon whiskey'.

Nonetheless, the US Congress's Senate recently declared September to be 'National Bourbon Heritage Month' and bourbon as 'America's Native Spirit'.

Whatever the merits of this official imprimatur, consumers clearly are very pleased with the product. Bourbon production has doubled since 1999, and exports jumped from $623 million to $713 million between 2006 and 2007 (14.4 per cent). Much of this growth has been fuelled by 'premiumization', the sales of the expensive, more refined whiskies. Sales of Woodford Reserve, for example, have increased at an average annual rate of 24 per cent over the past five years.

All American whiskey-makers, sensibly, are bringing more and more high-quality spirits to market. In addition to its age-old, black-labelled No.7 brand ($22), Jack Daniel's currently offers Gentleman Jack ($30) and Jack Daniel's Single Barrel ($50). Jim Beam, known for its white-labelled bourbon ($22) sells a whole line of top-shelf bourbons, including Knob Creek ($35), Baker's ($40), Basil Hayden's ($45) and Booker's ($60). Smaller distilleries also have exotic offerings. The Old Rip Van Winkle Distillery offers the twenty-year-old Pappy Van Winkle Family Reserve ($100).

And there appears to be a glimmer of hope for rye

lovers. For years, Old Overholt andJim Beam produced the only ryes to be found, and they were hard to locate. In the past decade, new ryes have arrived, including Rittenhouse ($15), Sazerac six-year-old ($25), Old Potrero Single Malt ($60) and Van Winkle Family Reserve 13 Year Old Rye ($75).

Chapter 6
The Twenty-first Century Whiskey World

The twenty-first century whiskey world is a postmodern world. What once was simple—grain made into hooch and sold mostly to local drinkers—has become mind-bogglingly complex. Whiskey has been globalized; there are hundreds of millions of potential customers with billions of dollars to spend on a range of whiskies and derivative whiskey products made around the globe. And for some, whiskey has ceased to be a simple product; it has become an object of adoration and worship.

Other Nations Enter the
Whiskey-Making Business

Scotland, Ireland, Canada and the US are not the only nations where whiskey is made. Over half a century ago, Japan leapt into the business of making single malt (*Shinguru moruto*) Scotch-like whiskies. Japanese brands have proliferated—Yamazaki, Hibiki, Hakushu, Yoichi and Taketsuru, to name a few, and the quality is often very high.

Likewise Australia. There is Bakery Hill of Victoria (founded in the late 1990s) and Lark Distillery of Tasmania (set up in the early 1990s), along with Great Southern Distilling Company, Hellyers Road Distillery and more.

If that was not enough, in recent years whiskey has been produced in the Czech Republic, Germany, New Zealand, Spain and Turkey. The Muree Brewery Co. Ltd of Rawalpindi, Pakistan produces eight- and twelve-year-old pot still, barley-based whiskies. In 1999, Mackmyra whiskey distillery cropped up 140 miles north of Stockholm, Sweden. It is producing barley-based whiskies that are peated and aged in Swedish oak casks.

Disturbingly, a company in Thailand marketed Cobra whiskey for a time, a rice-based spirit with ginseng, red peppers and a dead baby cobra in the bottle. Chris Carlsson of SpiritsReview.com described Cobra Whiskey as having a 'fishy, hot taste', and causing numbness in the mouth and 'tingling in the extremities'.

Brands Globalized, Whiskey Associations Unspun

A century or more ago, say, a man named McNaught might have owned the McNaught Whiskey Company, which owned the MacNaught Distillery and produced the McNaught Whiskey that was sold in local markets.

The evolution of globalization has unspun the simple

relationships between companies, distilleries, brands and markets. Wild Turkey Bourbon, an iconic American whiskey, is an example of this postmodern phenomenon.

The top of the Wild Turkey label reads, 'Estd. AN 1855,' with the words 'Austin Nichols' beneath it, and the words 'Wild Turkey' printed lower still. Some consumers might interpret this to mean that the Austin Nichols distillery has produced Wild Turkey whiskey since 1855, and that most of it is sold to Americans. In fact, the Wild Turkey Distillery in little Lawrenceburg, Kentucky (population 9,014) is owned by Italy's Gruppo Campari, which acquired it from France's Pernod Ricard in early 2009 for $575 million. Wild Turkey sells 2 million US gallons (7.5 million litres) in 40 nations around the world. And Austin Nichols? That's a vestige—1855 was the year that the Austin Nichols grocery company opened for business.

To take another example: where is the bar that serves the most Jameson Irish whiskey? Near the giant Midleton distillery in Cork, Ireland? No, the Local Irish Pub in Minneapolis, Minnesota, a state where less than 7 per cent of the population claim Irish heritage. The pub's customers consumed 1,600 US gallons (6,000 litres) in 2008, or about 22 bottles a day, mostly in the form of cocktails.

New Individualized, Derivative
and Associated Products

Globalization cuts both ways for whiskey-makers.

With more customers to serve, there are more profits to be found. But there is more competition than ever. No whiskey-maker can rest comfortably in the assumption that his brand will keep its place in public houses and shops. This anxious environment has fuelled the growth of new whiskies and whiskey-derived products made for increasingly specific markets.

Independent bottlings of whiskey have been around since the nineteenth century at least, when merchants bought whiskey and sold it under their own brands. Today, independents do not hide the provenance of the whiskey; they tout it. They aim to reach the consumer who wants something rare. So they buy whiskey from renowned producers, hold it, then bottle and sell it at unusual years and strengths. Take the $100 GlenTaite, for example. The Macallan markets many whiskies, but none at nineteen years. This particular Scotch can only be acquired by those individuals who shop at the Sam's Club warehouse stores in the US. There are many independent bottlers; Gordon & McPhail, Murray McDavid and Scott's are some of the well-known names. A newer entrant to the independent bottlings market, Wemyss Vintage Malts, employed Charles MacLean, the famed whisky writer, to pick casks. Wemyss's whiskies typically run $75 and more per bottle, or £25 or more in the UK.

As we have seen, the association of nations, ethnic identities and types of whiskey is as old as whiskey itself. Today, we still speak of Irish whiskey, and Scotch whisky, even if the companies that produce them are neither Irish nor Scottish.

Additional whiskey-identities have recently come on the scene. For a time, JBB (Greater Europe) PLC (now Whyte and Mackay Ltd) produced Hamashkeh whisky for Observant Jews. In order to ensure that this Scotch met the exacting standards required by the Kosher label, a Rabbi was present when the whisky was blended in Invergordon and when it was bottled in Grangemouth, Scotland. Something similar appeared across the pond—Old Williamsburg No.20 is a three-year-old bourbon. This kosher brand is named for the Brooklyn, New York neighbourhood where many Orthodox Jews live. More kosher whiskies may be on the way. Cooley Distillery in Ireland has received kosher certification, although the company has not yet released any kosher whiskies.

To the surprise of many, in 1996 Quentin Crisp Single Cask Whisky was released in the UK. Previously, few other distilled spirits makers had marketed their products towards homosexual consumers. Crisp Single Cask was the first whiskey to make such a move. The brand came and went in a blink, but it may prove to be a forerunner of more individualized and identity-specific whiskies.

Maker's Mark Bourbon, for example, has produced limited-edition bottles for the fans of the University of Kentucky basketball team and the Kentucky Derby, and for charitable groups. Those who visit the distillery can personalize their own Maker's Mark bottle by autographing and dating it, and then hand-dipping its neck in a vat of hot wax. One can easily imagine other whiskey-makers producing special edition whiskies for observers of particular holidays or special dates (we already have seen millennium-themed whiskies), and for enthusiasts for particular causes or pastimes.

Many distilleries sell whiskey futures to drinkers. Customers pay to claim barrels of ageing whiskey, and then may decide the date when the barrel is emptied and bottled. Some distillers, like the Ellensburg Distillery of Washington State, even put personalized labels on these bottles.

Whiskey purists often act as if the only way to enjoy whiskey is straight from the still or barrel. They often express contempt at the idea of mixing whiskey with anything else. But as we saw in chapter Two, some of the earliest mentions of whiskey described it as spirit mixed with herbs, honey and other items. The popularity of whiskey liqueurs has bloomed and faded over the years, and it may be spring season again. Long-time brands such as Drambuie Scotch liqueur, Irish Mist and Bailey's continue to hold shelf space in bars and stores, and new concoctions have arrived, such as Celtic Crossing, an Irish whiskey liqueur, and Wild Turkey's American Honey Liqueur. In 2009, Beam Global Spirits & Wine, Inc. brought out a bourbon, Red Stag, that is infused with black cherry flavours.

Similarly, whiskey-makers have had some success at selling mixed versions of their products. The notion is simple—save people the effort of mixing a cocktail by providing it in a bottle. Jack Daniel's may have been the first to attempt it on a large scale in the late 1980s, with its Jack Daniel's Country Cocktails (Lynchburg Lemonade, Black Jack Cola, etc.)

Whiskey, as a product, has leapt far beyond the bottle. Brands that have built fervent consumer bases have produced associated products. These go way beyond the cheap key chains handed out at concerts and dropped in gift bags, and the coasters and promotional barware in pubs. Whiskey-lovers actually pay, often a lot, for these associated products. These items are not produced to promote a brand; rather, they are designed to capitalize on the existent, intense consumer devotion to it. Jim Beam, the mega-bourbon producer, is an ace at this game. Its fans can buy Jim Beam sauces (barbeque, hot, marinade, steak and chicken-wing), in addition to Jim Beam Salsa and Jim Beam Beef Jerky. And that is just the food items. There also are Jim Beam clothes (T-shirts, boxer shorts, jackets) and homeware (café tables, directors' chairs, grill tools, wall-mounted, rotating pub lights for the home, and so on).

Adoration and Worship

In order to satisfy acolytes, whiskey temples have arisen around the world. The BlueGrass Tavern in Lexington, Kentucky serves 168 different brands of bourbon; the Cask in Tokyo has a large collection of extremely rare Scotch whisky, including ancient bottles of Black and White and Old Parr. Cadenhead's Whisky Shop in Edinburgh, Scotland sells over 200 Scotch whiskies; Binny's Beverage Depot in Chicago offers over 500 whiskies of all types. Whiskey distilleries around the world have kicked open their doors and built visitors' centres to receive the thirsty hordes.

In Graham Greene's 1940 novel *The Power and the Glory*, readers met the 'whiskey priest', a churchman whose cursed pride led him away from God and into debauchery. Today's whiskey priests are more in the cast of Virgil in Dante's *Inferno*. Writers like Charles Cowdery, John Hansell, Charles Maclean, Jim Murray, Gary Regan and Gavin Smith teach those who are new to the whiskey world the whats, whys and hows of whiskey. Their writings appear in magazines such as *Malt Advocate*, *Whisky Magazine*, the *Bourbon Country Reader*, the *Whiskey Life* and numerous online publications. Drinkers gobble up their books and often flock to festivals to hear these priests hold forth on whiskey.

To further feed the passion of the world's whiskeyheads, there now are droves of whiskey events and experiences to be had. Pubs hold tastings, some of the aforementioned whiskey priests conduct private samplings, and there are gatherings in North America and Europe, like the Single Malt & Scotch Whisky Extravaganza, Whiskey Fest and Whiskey Live, that attract hundreds at a time. Over 1.2 million people visit Scotland's distilleries each year, and some 55,000 visitors from more than a dozen countries head to Bardstown, Kentucky (population: 11,500) for the annual Kentucky

Bourbon Festival.

Today, whiskies still can be found for $10 or £15 a litre, but they have been shoved aside by legion luxury whiskies. First it was the single-malt Scotches, then the blends and now Irish, bourbon and rye whiskies appear in sleek bottles and are priced at $40 a bottle (or £30) or more.

Auctions now are held to sell of rare lots of whiskies, and the prices often defy belief. Nuns' Island distillery of Galway, Ireland closed in 1913. When an unopened bottle of its whiskey surfaced a few years ago, bidding was to begin at $150,000. Distilleries have taken notice of this trend, and have been digging up more and more ancient whiskies from their warehouses. Dalmore has offered a 50-year-old Scotch packaged in a crystal decanter for $12,000, and The Macallan's Fine & Rare 1926 fetches $38,000 a bottle. To titillate readers, whiskey publications regularly carry reports of eye-popping prices and photographs of these super-rare products, a sort of 'whiskey porn' that excites readers, the majority of whom could never afford or acquire such bottles.

But there are those lucky few people who have the wallets. Two individuals on opposite sides of the world have become famous on this count. Valentino Zagatti of Lugo Di Ramagna, Italy, has a Scotch collection numbering in the thousands, and dating back to the late nineteenth century. His collection is so impressive that Italy's Formagrafica publishing company has issued two coffee-table sized picture books of Zagatti's holdings, *The Best Collection of Malt Scotch Whisky* (1999) and *The Best Collection of Malt Part 2: Whiskies and Whiskies* (2004).

Then there is the man known throughout the English-speaking whiskey world as Harvey. Purportedly an heir of an industrial fortune, this whiskey-lover has lined the walls of multiple rooms of his hulking Washington, DC, townhouse with bottles. Crates of new acquisitions arrive every few days, which Harvey painstakingly catalogues, tastes and rates. The value of the collection is inestimable. Once when I visited Harvey, I tried to stump him by asking for a taste of an uncommon brand of Scotch. He led me from room to room, and within minutes we had fourteen different versions of that particular whiskey. He generously offered to get more of them, but suggested (quite rightly) that tasting them all on the same night might be overkill.

Among the hard-core whiskey fans, a small but growing purist, and even backward-looking, backlash has developed. It disdains much of what is called whiskey today and demands something less polished and more 'authentic'. One can see it in the rise in sales of whiskies offered at 'cask strength', or straight from the barrel without any watering down to 80 proof or filtering.

One also detects it in the rapid rise of whiskey micro-distilling. Much like the micro-beer movement that began in the late 1970s, micro-distillers tout their small production capacities, their use of locally grown and organic grain and the use of pot stills and 'old world' craftmanship. The USA has been a hotbed of this activity. Producers include Belmont and Wasmund's of Virginia, Charbay and St George's of

California, McCarthy's of Oregon, Strahan's of Colorado and Tuthill of New York.

Astonishingly, this back-to-our roots movement has helped bring Irish poteen back to the market. The path was led by Bunratty Mead & Liqueur Company (County Clare) and Knockeen Hills (County Waterford). Both of these poteens are unaged and water-clear. Bunratty weighs in at 80 proof, while Knockeen Hills is produced at three strengths: 120 proof, 140 proof and an eye-popping 180 proof. Sales are modest, but climbing. Not much more than a century ago, the Crown employed armed troops to quash the production of poteen. In a surreal twist, now one finds bottles of Knockeen Hills for sale in Heathrow Airport's Terminal 3 .

Whiskey-makers and public relations firms like to speak of the 'good old days of whiskey'. If this book has shown anything, it is that whiskey drinkers never have had it better. Through sensible government action, and through capitalist competition, whiskey has become better than ever. Consumers around the globe have access to more and better types of whiskey. A woman visiting a store in South Africa can choose the brand of Scotch she wishes to purchase for serving to guests visiting her home. A man sitting in a bar in Brazil can use his laptop computer to buy a friend sitting in a Chicago eatery an Irish Coffee via the website GiveReal.com. A group of friends in Germany can arrange to purchase a cask of their own from a Scottish distillery.

For this rich bounty of choices, we should raise a toast— to whiskey, and to the good life!

Recipes

Hot Toddy

Pour 1 to 3 fl. oz. (30–90 ml) of any whiskey in a large mug. Add a lemon slice, as much honey as you please and 5 or more fl. oz. (150 ml) of boiling water.

Irish Coffee

Pour 1 to 2 fl. oz. (30–60 ml) of Irish whiskey in a mug or heat-resistant glass, add as much sugar as you please and 6 fl. oz. (180 ml) of hot coffee. Float cold cream or whipped cream on top.

Manhattan/Rob Roy

Pour 2 fl. oz. (60 ml) of bourbon or rye whiskey and 1 fl. oz. (30 ml) of Sweet Vermouth in a cocktail shaker that is one-third full of ice. Add 1 to 3 dashes of Angostura Bitters. Shake and strain the mixture into a glass. Drop a maraschino cherry in it. For a Rob Roy cocktail, substitute Scotch.

Mint Julep

Put 3 or more fresh mint leaves in a sturdy glass, and

pour in 3 to 5 fl. oz. (90–150 ml) of simple syrup (which is a 50/50 mixture of refined white sugar and water). Pestle the mint leaves until they start to tear. Pour 2 to 3 fl. oz. (60–90 ml) of American whiskey or bourbon in the glass, stir, and fill the glass with ice (preferably shaved or crushed). Garnish with a sprig of mint.

Scotch and Soda

Pour 1 to 2 fl. oz. (30–60 ml) of Scotch and 6 fl. oz. (180 ml) of Club Soda in a glass with a few ice cubes. Garnish with a twist of lemon if you please.

Whiskey and Cola

Pour 1 to 2 fl. oz. (30–60 ml) of American, bourbon or Tennessee whiskey and 6 fl. oz. of cola in a glass with a few ice cubes. Garnish with a slice of lemon or lime if you please.

Whiskey Sour

Pour 2 fl. oz. (60 ml) of bourbon or rye whiskey in a cocktail shaker that is one-third full of ice. Add ½ fl. oz. (15 ml) of lemon juice and ½ fl. oz. (15 ml) of simple syrup (which is a 50/50 mixture of refined white sugar and water). Shake and strain the mixture into a glass with or without a few ice cubes. Garnish with a maraschino cherry and a slice of orange.

Some Widely Available, Recommended Brands

American/Bourbon/Tennessee Whiskies

Buffalo Trace Kentucky Straight Bourbon Whiskey

George Dickel #^{12} Tennessee Whisky

Old Forester Kentucky Straight Bourbon Whisky

W. L. Weller Special Reserve Kentucky Straight
Bourbon Whiskey

Canadian Whiskies

Alberta Premium Rye Whisky

Canadian Club Whisky

Crown Royal Special Reserve Blended Canadian Whisky

Irish Whiskies

Knappogue Castle Single Malt Irish Whiskey 1994

Redbreast Irish Whiskey

The Tyrconnell Single Malt Scotch Whiskey

Jameson Irish Whiskey

Rye Whiskies

Jim Beam Rye 1 Whiskey

Old Overholt Rye Whiskey

Wild Turkey Rye Whiskey

Scotch–Blended and Vatted

Bell's Finest Old Scotch Whisky

Chivas Regal

Compass Box Hedonism Scotch Whisky

Johnnie Walker Black Label Blended Scotch Whisky

Johnnie Walker Green Label Scotch Whisky

Single Grain Whiskey

Greenore Irish Whiskey

Scotch–Single Malt

Ardbeg 10 Year Old Single Malt Scotch Whisky

Glenlivet 12 Year Single Malt Scotch Whisky

Laphroaig 10 Year Old Single Malt Scotch Whisky

Old Fettercairn Highland Single Malt Whisky 10 Year

The Glenrothes Single Malt Whisky 1994

Select Bibliography

Cowdery, Charles, *Bourbon Straight: The Uncut and Unfiltered Story of American Whiskey* (Chicago, IL, 2004)

Crowgey, Henry G., *Kentucky Bourbon: The Early Years of Whiske-making* (Lexington, KY, 1971)

Dabney, Joseph E., *Mountain Spirits: A Chronicle of Corn Whiskey from King James' Ulster Plantation to America's Appalachian and the Moonshine Life* (New York, 1974)

Daiches, David, *Scotch Whisky: Its Past and Present* (London, 1969)

Forbes, R. J., *A Short History of the Art of Distillation*, 2nd edn (Leiden, Netherlands, 1970)

Hume, John R., and Michael S. Moss, *The Making of Scotch Whisky: A History of the Scotch Whisky Distilling Industry* (Edinburgh, 2000)

Jefford, Andrew, *Peat Smoke and Spirit: The Story of Islay and its Whiskies* (London, 2004)

MacClean, Charles, *Malt Whisky* (London, 2002)

Maguire, E. B., *Irish Whiskey: A History of Distilling, the Spirits Trade and Excise Controls in Ireland* (New York, 1973)

Mulryan, Peter, *The Whiskeys of Ireland (Dublin, 2002)*

Murray, Jim, *Jim Murray's Whisky Bible 2006* (London, 2005)

Regan, Gary, and Mardee Haiden Regan, *The Book of Bourbon and Other Fine American Whiskeys* (Shelburne, VT, 1995)

Smith, Gavin D., *The A to Z of Whisky* (Glasgow, 1993)

—, *The Secret Still: Scotland's Clandestine Whisky Makers* (Edinburgh, 2002)

Steadman, Ralph, *Still Life with Bottle: Whisky According to Ralph Steadman* (London, 1994)

Whiskey Cookbooks

McConachie, Sheila, and Graham Harvey, *The Whisky Kitchen* (Thatcham, Berkshire, 2008)

O'Connor, Aisling, *Cooking with Irish Whiskey* (London, 1995)

Regan, Gary, and Mardee Haiden Regan, *The Book of Bourbon and Other Fine American Whiskeys* (Shelburne, VT, 1995), pp. 293–331

Tolley, Lynne, *Jack Daniel's Spirit of Tennessee Cookbook* (Nashville, TN, 2009)

Websites and Associations

AlcoholReviews.com

The Chuck Cowdery Blog (on whiskey in the US)
http://chuckcowdery.blogspot.com

IrishWhiskeyNotes.com

IrishWhiskeySociety.com

The Kentucky Bourbon Trail
www.kybourbontrail.com

Malt Advocate Magazine
www.maltadvocate.com

Oscar Getz Museum of Whiskey
www.whiskeymuseum.com

The Scotch Whisky Heritage Centre
www.scotchwhiskyexperience.co.uk

The Scotch Malt Whisky Society

www.smws.com

SpiritsReview.com

StraightBourbon.com

Whisky Magazine (FR)

www.whisky.fr

Whisky Magazine (UK)

www.whiskymag.com

The Whiskey Portal

www.whiskyportal.com

Acknowledgements

I am indebted to Andrew F. Smith, editor of Reaktion's *Edible* series, for giving me the opportunity to write this book, and for being a source of inspiration.

Hearty thanks go to Reaktion Books' Michael Leaman and Martha Jay for their eagle-eyed, keen-minded editorial assistance, and to Dr R. Sam Garrett of the Library of Congress, who lent an unbendable listening ear to me as I worked my way through this book.

I also owe thanks to the following for their help with data, illustrations and materials for this book: Richard Anthony, freelance photographer; Elizabeth Balduino of Taylor PR; Jennifer Bowling of the Winery Exchange; Chris Carlsson of SpiritsReview.com; Robin Coupar of Skyy; Chris Huey and Walt Tressler, Brown-Forman Corporation; Liske Larsen of Cutty Sark International; Marianne Martin of the Colonial Williamsburg Foundation; Jayne Murphy and Caroline Begley of Irish distillers Pernod Ricard; Susie Rea of Smarts communications; Jack Teeling and Jennifer Grainger of Cooley Distillery; Katie Young and Chelsea

Cummings of Qorvis Communications; and the unbelievably helpful employees of the Prints and Photographs Division of the Library of Congress.

Any shortcomings or gaffes in this book are the responsibility of the author.

Photo Acknowledgements

The author and publishers wish to express their thanks to the below sources of illustrative material and/or permission to reproduce it.

Photos Richard F. Anthony: pp. 007, 016, 063, 066; photo BrownForman Corporation: p. 111; from Hieronymus Brunschwig, *Liber de Arte Distillandi* (Strasbourg, 1512): p. 035; Bushmills Irish Whiskey: p. 090; City of Toronto Archives: p. 116; photos Cooley Distillery: pp. 088, 093; photo Mike DiNivo/Action Sports Photography: p. 130; photos Dave Doody and Tom Green/The Colonial Williamsburg Foundation: p. 038, 039, 136; photo The Edrington Group: p. 061; photo Tim Eubanks/Sam's Club: p. 129; from Harrison Hall, *The Distiller* (Philadelphia, 1818): p. 101; photo John Hawkins: p. 125 top; after John R. Hume and Michael S. Moss, *The Making of Scotch Whisky: A History of the Scotch Whiskey Distilling Industry* (Edinburgh: Canongate Books, Ltd, 2000): pp. 052, 053 top, 053 bottom, 085; photo Irish Distillers Limited: p. 086; photos Kevin R. Kosar: pp. 009, 011, 020, 046; photos Library of Congress, Washington,

DC: pp. 003, 013 bottom, 045, 051, 057, 074, 075, 078, 098, 102, 104, 107 top, 107 bottom, 108, 112, 114, 117; from Ian MacDonald, *Smuggling in the Highlands* (Inverness, 1914): p. 049; photo Makers Mark: p. 019; from Petrus Andreas Matthiolus, *Opera quae Extant Omnia* (Frankfort, 1586): p. 036; from John L. Stoddard, *John L. Stoddard's Lectures*, Suppl. vol.1, *Ireland* (Boston, 1901): p. 077; photo Johnnie Walker: p. 058; photos Scotch Whisky Association: pp. 004, 017, 053 bottom; photo Smithsonian National Postal Museum Collection: p. 119; photo Spirits-Reviews.com/ Chris Carlsson: p. 125 bottom; from Sydney Young, *Distillation Principles and Processes* (London, 1922): pp. 013 top, 014, 032.

图书在版编目（CIP）数据

威士忌 / （美）凯文·R.科萨著；吴德煊译.
-- 北京：北京联合出版公司，2024.4
（食物小传）
ISBN 978-7-5596-7393-0

Ⅰ．①威… Ⅱ．①凯… ②吴… Ⅲ．①威士忌酒-
普及读物 Ⅳ．① TS262.3-49

中国国家版本馆 CIP 数据核字（2024）第 023637 号

威士忌

作　　者：〔美国〕凯文·R.科萨
译　　者：吴德煊
出 品 人：赵红仕
责任编辑：孙志文
产品经理：夏家惠
装帧设计：鹏飞艺术
封面插画：〔印度尼西亚〕亚尼·哈姆迪

北京联合出版公司出版
（北京市西城区德外大街 83 号楼 9 层　　100088）
北京天恒嘉业印刷有限公司印刷　　新华书店经销
字数 117 千字　889 毫米 ×1194 毫米　1/32　9 印张
2024 年 4 月第 1 版　　2024 年 4 月第 1 次印刷
ISBN 978-7-5596-7393-0
定价：59.80 元

版权所有 侵权必究
北京市版权局著作权合同登记　图字：01-2022-5540 号